COMMUNICATIONS SATELLITES

Communications Satellites

Proceedings of a Symposium
held in London on 12 May 1961

Organized by the
British Interplanetary Society

Advisory Editor: L. J. CARTER

The British Interplanetary Society

1962

ACADEMIC PRESS
LONDON AND NEW YORK

ACADEMIC PRESS INC. (LONDON) LTD.

BERKELEY SQUARE HOUSE

BERKELEY SQUARE

LONDON, W.1

U.S. Edition published by

ACADEMIC PRESS INC.

111 FIFTH AVENUE

NEW YORK 3, NEW YORK

Library of Congress Catalog Card Number: 61-18804

Printed in Great Britain by W. & J. Mackay & Co. Ltd., Fair Row, Chatham

LIST OF CONTRIBUTORS

BRADLEY, J., British Aircraft Corporation, Ltd., Luton, Bedfordshire, England.

HAVILAND, R. P., General Electric Company, Missile and Space Vehicle Department, Philadelphia, Pennsylvania, U.S.A.

HEBENSTREIT, W. B., Space Technology Laboratories, Inc., Los Angeles, California, U.S.A.

HILTON, W. F., 5 Grange Avenue, Twickenham, Middlesex, England.

MUELLER, G. E., Space Technology Laboratories, Inc., Los Angeles, California, U.S.A.

PARDOE, G. K. C., The de Havilland Aircraft Co. Ltd., London, England.

PIERCE, J. R., Bell Telephone Laboratories, Murray Hill, New Jersey, U.S.A.

SANDEMAN, E. K., British Aircraft Corporation, Ltd., Luton, Bedfordshire, England.

SENN, G., U.S. Army Signal Research and Development Laboratory, Fort Monmouth, New Jersey, U.S.A.

SIGLIN, P. W., U.S. Army Signal Research and Development Laboratory, Fort Monmouth, New Jersey, U.S.A.

SPANGLER, E. R., Space Technology Laboratories, Inc., Los Angeles, California, U.S.A.

FOREWORD

By W. R. Maxwell

President of the British Interplanetary Society

All activities in space are expensive and for this reason little can be undertaken in this field without financial support from Governments. Yet it is difficult to specify in anything but rather general terms the likely benefits which might be derived from astronautical ventures, and the arguments which can be put forward for spending considerable sums of money on them do not in general appeal to those who hold the purse strings. The communications satellite, however, is a spatial undertaking which can be argued on a purely financial basis, and because of its commercial possibilities either as a replacement or a supplement to other forms of communication such as the telephone, has attracted a great deal of attention and study during the last year or so. Many different kinds of communications satellite systems have been put forward which vary according to whether they use active or passive satellites, the area of the earth to be covered, the type of orbit preferred, the best type of coding and decoding system to use, and so on. There has also been much argument and discussion on the estimates of system costs, the extent of likely usage, allocation of wavelengths and other practical considerations. The time delay and echo effect with a communications satellite system has likewise received considerable attention and its importance or otherwise debated. In view of all this activity the British Interplanetary Society thought it timely to hold a symposium on the subject at which papers could be given by authorities in the field and the various practical and technical considerations debated. The symposium was held on the 12th May 1961, and the various papers submitted, which include several from the U.S.A., have been collected together in this book and give an up-to-date account of current thinking on many aspects of the subject.

FOREWORD

by W. R. Maxwell

President of the British Interplanetary Society

CONTENTS

CONTENTS

THE ENGINEERING AND ECONOMICS
OF SATELLITE COMMUNICATIONS SYSTEMS

G. K. C. PARDOE

The de Havilland Aircraft Co. Ltd., London, England

I. INTRODUCTION

It is particularly fitting that the B.I.S. should have organized a Symposium on Communications Satellites, because it was Arthur Clarke, a past president of our Society, who first drew attention to the possibilities of using satellites around the earth as radio repeater stations in a global telecommunications system. This was 16 years ago in the pre-Sputnik era, in the immediate post-war period when rockets were not quite "decent". Almost 10 years elapsed before interest was re-kindled by Dr Pierce and now of course during the last 5 or 6 years we have seen numerous papers presented dealing with the many and varied aspects of communications satellite systems. Here we shall continue the theme a little further.

The vast majority of these papers have concentrated on the purely theoretical and technical aspects and only occasionally have the economics been discussed. In the early days this was understandable, and did not much matter; but now that prototype operational communications satellites are fast becoming realities, it is important that the system be viewed as a whole. The subject is probably unique in so far that no single question relative to any single aspect of the system can be answered out of context of the whole. This is particularly true in the all important matter of cost.

I thought then that it might be of interest first of all to review the factors—all of the factors—that influence the engineering and economics of a satellite system, and the manner in which they do it; and then to consider certain of those factors which exert the greatest influence on the problem.

I do not intend to reach any startling conclusion during this paper, but rather to point out how many things must be considered before such conclusions are reached. To those of you who are already expert in this matter of space communications systems, I ask your indulgence for presenting a paper of a broad descriptive nature rather than with a detailed technical content. However, to those of you who may not be quite so familiar with these matters, I felt that such a paper as this may well set the scene for the more particular aspects which will be dealt with in later papers.

II. Overall System

Figure 1 shows the main elements of a communications satellite system. It shows how each of the 4 primary elements of the system are connected and interconnected. The central position adopted for the factors of cost and choice of orbit pattern is indicative of their overall importance, besides showing that they are common parameters.

When a system operator or a telephone administration considers the installation of a new service, he presumably has in mind the areas he wishes to serve and an estimate of the potential demand for the type of

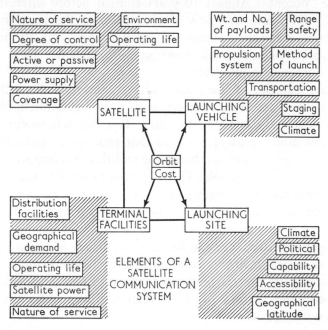

Fig. 1

service he proposes to introduce. Thus, type of service, continuity of service, coverage and number of channels are among the first considerations to be made. The availability or otherwise of terminal facilities and distribution networks come next and then climate and length of operational life. The latter two primarily affecting cost by virtue of having to design and engineer equipment for long term reliability. Last, but by no means least, is the anticipated transmitter power and the receiver sensitivity within the satellite. These directly affect the size and cost of the ground aerials, reflectors, transmitters and receivers. Low satellite transmitter power demands large aerial reflectors. The cost of such aerials, for instance, is closely related to their size; it would appear to increase roughly in proportion to the square of the dish diameter.

In the satellite quadrant we note again the direct effect that the orbit pattern and the number of satellites makes on satellite design. Special note should perhaps be made concerning the decision to use active or passive repeaters—passive satellites having an especial significance when considering the ground station aerials, transmitters and receivers. The fact that the satellite does not amplify the ground transmitted signal means that the size of ground aerial reflectors increase accordingly and we have just seen the effect of size on cost.

The dependence of satellite design on orbit pattern can be summed up by the mnemonic ACTEW—the greater the Altitude, the greater the Coverage, therefore the greater the Transmitter power required, therefore more Electric power, therefore more Weight; or conversely the lesser the altitude etc.

This brings us straight into the launching vehicle region. Here it will be seen that the total weight to be placed in orbit is the all important criterion. This weight directly determines the size and number of stages of the vehicle, after due consideration has been given to the method of propulsion. Having thus arrived at a given size and weight of vehicle, the method and problems of launching—range safety measures—and transportation of the vehicle during manufacture and test all begin to exert their influence on design. Before this can be settled even provisionally, though, an attempt must be made to specify and locate the launching site, and here we enter the last quadrant of this diagram; a most important region because it brings with it considerations of a political nature, geographical location, national security, range safety and accessibility. The fundamental requirements are inherently incompatible. From a safety point of view it needs, ideally, to be in the centre of an isolated, sparsely populated area of some hundreds of thousands of square miles in extent. From an operational and support point of view it needs, ideally, to be easily accessible and surrounded by technical and industrial back-up facilities. The most important factor though is the geographic latitude of the site—this directly determines the limits of orbit inclination and here we see the close of the system loop—orbit inclination influencing satellite coverage which in turn influences or determines service areas for the communications system.

Having thus covered, quickly, the overall system we can turn now to consider the individual regions in a little more detail. I propose to deal with them in the reverse order.

III. Launching Site

As we have just seen, geo-political considerations determine very largely the broad location of a launching site, particularly since most of the current generation space launching vehicles use modified military missiles as first stage boosters. This introduces a natural predisposition to launch from military bases and establishments, with all that this implies in the way of existing facilities—not forgetting, of course, the restrictions brought about for reasons of national security.

The size of launching vehicle can, alone, influence the method of launching. It is often commonly taken for granted that all space vehicle launchings will be made from ground sites, but this is not true. Already, in the case of space probes, launchings have been made from balloons, whilst launchings from mobile platforms or tenders at sea are under serious consideration for the very large booster rockets such as SATURN. Such launchings at sea avoid the cost of the elaborate precautions which have to be made at ground sites to overcome the risks and results of fire and explosion.

The latitude of the site north or south of the equator determines the minimum angle of inclination (to the equator) which can be achieved with a simple orbital injection technique. Thus satellites launched by rocket from say Woomera or Colomb Bechar, both at approximately 31° latitude (south and north respectively) can be placed into orbits between 31° and 90° (i.e. polar). Orbit inclinations between say 31° and 0° (equatorial) need further vehicle stages to accomplish the additional manoeuvres involved in changing plane. This not only adds complexity to the vehicle but it also results in a smaller payload for a given first stage booster. All of this is reflected in the cost; not only does the direct cost go up as a result of the elaboration in manufacture, launching, tracking, command etc., but there is a corresponding rise in the basic cost per lb mass of payload in orbit. This latter ratio being a useful indicator or measure of space vehicle efficiency.

One might say that given a free hand the natural choice for a launching site would be for one that is situated on the equator—but, as we have previously noted, most of the world's rocket bases have been established elsewhere, as a result of other predominant requirements. But it is of interest to note that Hughes in their proposal for a lightweight communications satellite in equatorial 24-hr synchronous orbit specified Christmas Island as the most appropriate site for launching. This is an island in the middle of the Pacific something less than 2° north of the equator. The fact that the rocket they propose is a modified SCOUT with additional upper stages, simplifies the problem of supplies and facilities; but it must be noted that the cost of establishing even these facilities will have inevitable repercussions on the overall cost of the system.

A further point arising from the site location is its possible effect on the staging arrangements of the launching vehicle. Ideally the vehicle would be designed to achieve maximum velocity increments at each stage—a problem of dynamics involving the optimization of staging mass ratios—and the total weight and impulse from propellents in each stage. This is relatively simple whilst it is unaffected by outside considerations; but to achieve a given orbit pattern from a given launch site it is necessary to launch the vehicle in a given direction. It can and so often does happen that when the satellite flight plan is superimposed on the ground map it is found that our ideal staging gives rise to political headaches. We find, for instance, the trajectories of the expended first stages terminating in densely populated areas or in foreign territories. This latter is undesirable even when the impact area is virtually uninhabited; witness the political repercussions when pieces of THOR fell on Cuba and parts of a lunar probe landed in South Africa.

Having thus decided upon a site which is acceptable politically and geographically; and having, probably, tailored the vehicle staging to suit, we find that the site is situated in the middle of an arid zone or marshy waste. It may so happen that the military have already put in a lot of time and money reclaiming land and making the site amenable for human occupation, coupled with the building of roads, railways and airstrips, in which case the cost of launching could be a simple fee—

albeit a stiff one! But it could so happen that the site must be developed from nothing; in which case the total cost will appear in the bill for the overall communications system, and this, in turn, may force the systems engineer to reconsider his previous decisions.

This process of weighing and re-weighing the pros and cons between pairs of variables out of a multiplicity of dependent variables is a continuous one—but as I intend to show later it is often simplified in practice by predetermination of some of the variables—making them in fact constants.

IV. The Launching Vehicle

Ideally one would start considering the launching vehicle from the payload point of view. However, as is the case throughout the system no one aspect is independent and, due to existing design limitations, we must "cut our coat according to our cloth". The satellites, therefore, tend to have weight and size limitations imposed on them.

As is well known, suitable fuel is the limiting factor, since too much of the vehicle is taken up with fuel storage. The ideal fuel would have high specific impulse coupled with high density thus permitting a small and reliable structure. From the handling aspect, solid fuels are more convenient than liquid but the associated control problems and the difficulty of ensuring even burning together with their low specific impulse indicate that, for a long time to come, liquid fuels will be employed where reasonable payloads are required. This will call for elegant design concepts for the structure since a high strength to mass ratio will be required. Existing techniques have followed the line of making the fuel tank the skin of the vehicle and using gas pressurization to maintain structural rigidity. Fuel is fed from the tanks to the motors by means of turbo-pumps.

Staging, although introducing complexity, enables the vehicle to shed unwanted structural weight at a point in the trajectory where the majority of the propulsion system is redundant. This process may usefully be repeated several times (e.g. SCOUT, which has 4 stages). It is most advantageous to employ fuels with maximum specific impulse in the upper stages where higher velocity increments are required. Range safety requirements will dictate the points at which it will be permissible to shed the earlier stages and this will in all probability reduce the efficiency of the staging arrangements.

Since the vehicle considered will probably be launched from a position remote from its factory, the problem of transporting such a bulky object must be considered at the design stage. The individual stages will obviously travel separately with no dismantling problems. We are still left, however, with the first stage. In certain instances the propulsion unit could be detachable from the tank section which could then travel overland on its own trailer or by river on specially constructed barges or by sea as deck cargo.

In addition to the staging arrangements dictated by range safety requirements it will be necessary to provide some form of self-destruction for the vehicle; so that in the event of a guidance failure, a suitable signal could be sent from the ground to destroy the vehicle.

The degree of precision dictated by orbital requirements will affect the design of the guidance system. It would seem that internal guidance arrangements will need to be supplanted by ground radar plotting of the vehicle trajectory. Initiation of upper stages by radio command would simplify the vehicle electronics whilst providing some degree of adjustment to the orbit.

V. The Satellite

The satellite of course is the vital part of the system—once the means is available to get any reasonably sized payload into orbit!

The type of service required is the obvious starting point for consideration of the satellite design; it seems perfectly clear that by the end of this decade or so, we shall have satellites providing every conceivable type of communication service, ranging from telephone, telegraphic, television and data transmission etc. The question is, really, "What is the order in which these services are to be opened up?" Telephone facilities are at present available by alternative means, whereas live, intercontinental television can only be accomplished by means of communications satellite systems. Obviously the desirable thing to do is to create a satellite which can have both telephone and television signal capacity. The bandwidths and associated transmitter power necessary for this seem reasonable even for the first few satellites going into a system. Active or passive types of satellites are clearly related to the type of service, as well as the operating life or reliability aspects; but there are of course one or two clear-cut issues. For example, a delayed repeater satellite can only be active. There is of course a further sub-division. This one might refer to as a "double active" satellite, in other words a satellite not only equipped with the means of receiving, amplifying, and transmitting the signals back to the ground, but also equipped with separate antennas, beaming signals to adjacent satellites in a synchronous network; this satellite would then either repeat them to other satellites, or dump them to ground stations. Such a system appears to have many disadvantages from the point of view of complexity of aerials, attitude stabilization, size, weight etc., and in any case intermediate ground stations must be established for linking into local ground networks. It may well be that these double active satellites will find their best use in military communications, where indeed reliance should not be based on intermediate ground stations, and furthermore the narrow beamwidth for the inter-satellite transmissions provides a system which cannot be intercepted or jammed other than by direct interference with each satellite in the system. We may well see such a double-active system being used therefore in the military field rather than the commercial field.

The capacity of the satellite to handle signals has already been touched upon: the generation of electrical power is of course of vital significance to the active satellite and it would seem that much more information must be acquired on the radiation hazards for the medium altitude orbits before one can plan with certainty on the use of solar cells and other electronics for life terms measured in many years. However, with other electrical generation systems already being worked on, it may well be that these radiation hazards will not have such a measure of influence on the choice of orbit heights as may be interpreted from certain evidence at the moment available.

The question of whether the satellite needs to be attitude stabilized in association with its directional antenna is a major issue, and a requirement to attitude stabilize a satellite causes a major weight penalty as well as complexity and unreliability. Elimination of this particular requirement would pay handsome dividends.

Another important factor worth touching on is that of the orbit pattern; in particular, whether synchronous position-keeping satellites should be established; or whether it be accepted that satellites appear at random. The launching capability influences this, since random patterns involving say 50 satellites only become an economic proposition if several satellites can be carried aloft in one launching vehicle. Another factor which could be somewhat disturbing in the random system, is the probability of interrupted service—even with (shall we say) a 98% probability of continuous service, one feels that the remaining 2% discontinuity could be embarrassing. Its occurrence could not necessarily be predicted and in any event it would be most disconcerting if it occurred in the middle of a vital piece of data transmission; or during a telephone conversation, or a television programme for example.

VI. THE TERMINAL FACILITIES

As with the whole system, the terminal facilities are largely determined by customer demand. The type of link between any two positions is influenced by several factors, the predominant ones being:

(1) type of service required;
(2) number of channels;
(3) continuity of service;
(4) distance apart;
(5) terrain to be covered; and
(6) cost.

Considering, for the moment, the purely geographical factors, certain primary points arise. A satellite link provides a means of long distance communication which can be as reliable as a cable link but without the installation problems associated with the latter. This advantage is obviously not restricted to transoceanic links since the provision, or introduction, of overland transcontinental links may be made impracticable by the terrain encountered. Two excellent examples of such a problem are to be seen in Africa and South America. In the absence of a cross-country link, telephonic communications between Accra in West Africa and Nairobi in East Africa must at present be routed via London. Similarly Lima in Peru has to be linked to Rio de Janeiro via Barbados. In both cases direct ground cables or ground microwave links are out of the question, but even if the problem of installation could be overcome, maintenance would prove extremely difficult and expensive. Furthermore, a sophisticated country like the United States of America could well decide to use a satellite link for transcontinental communications. The cost of a call via the satellite would compare most favourably with the costs of existing ground links. Naturally, in this country the scale of operation is much reduced and such a scheme would not be economical.

Other factors affecting the position of the terminal are the necessity for a comparatively local demand (to avoid excessive ground routing charges) together with suitable linkage facilities to the local distribution system.

The type of service required affects the design of the entire system since, obviously, the requirements for television transmission differ considerably from those relating to telephony or telegraphy. Initially the emphasis will undoubtedly be on telephony and telegraphy transmission but the desirability of a reliable intercontinental television link is obvious. The number of channels provided is initially determined by the customer (who may well be a national authority) but in certain instances this number may be subject to the limitation of satellite design. Another vital factor in the design of the whole system is the continuity of coverage required. As the orbit altitudes decrease it is necessary to increase the number of satellites in order to obtain more nearly continuous operation, the degree of which is determined by customer demand. This point will no doubt be considered in more detail by later speakers. Where such continuity is not desired and "real time" is not a criterion a satellite system such as COURIER can be considered.

The transmitter and receiver design of the satellite and terminal station cannot be considered separately. For obvious reasons the complexity of design and weight of components in the satellite must be kept to a minimum and this will result in increased complexity on the ground. Frequency limitations dictate an operating frequency in the kilomegacycle range and coupled with the low transmitter power in the satellite a large dish (up to say 100 ft diameter) for receiver and a smaller dish for the transmitter will be needed. Since one set of aerials will be tracking the satellite in use, a second set will be required to acquire the next satellite to be used as it comes into view. Since, however, links will undoubtedly be in more than one direction one can envisage a typical station having several pairs of aerials. An exceedingly low noise input stage such as a maser or parametric amplifier will be required since the satellite transmitted power will be in the region of a few watts.

In order to acquire the next satellite, the aerials will need to know in which direction to look (remember we are dealing with large dishes which will take several minutes to scan 180°). The solution would appear to lie in orbit prediction by means of computer using information derived from conventional tracking methods.

VII. General Comments

Throughout the consideration so far, of the various elements in this whole system, the deviations from the ideal conditions imposed by the realities of life have been strongly emphasized. It is therefore of interest to go back to the breakdown of the four main areas and consider what overall system limitations are imposed if one either chooses or is forced to take advantage of techniques or facilities currently available, as opposed to designing *ab initio*. For the purpose of demonstration I have taken the proposed BLUE STREAK satellite launcher and its facilities to establish this point.

1. Launching Site (Fig. 2)

The launching site being considered for this rocket is of course Woomera, Australia, and in some degree or another all of the elements of this part of the system are influenced or dictated by virtue of the

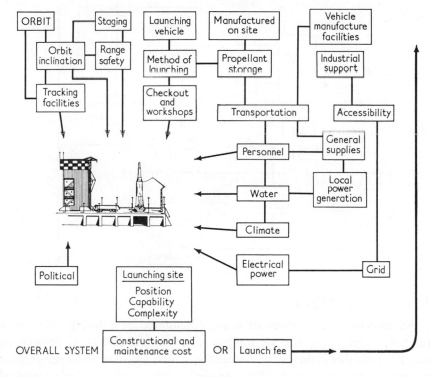

Fig. 2

existence of this particular launching site in an advanced degree of preparation.

2. *Launching Vehicle* (*Fig. 3*)

By the same token, most of the elements in the launching vehicle are influenced by the choice of BLUE STREAK as a first stage; in the sense that, within techniques most likely available in the foreseeable future, there is a limit to the load which the basic BLUE STREAK can lift off the pad. This, therefore, influences the size of the upper stages, although it should be noted that there will be a considerable extension of performance available using multi-stage vehicles based on BLUE STREAK. Enough to say therefore that the choice of booster certainly influences the whole of the quadrant containing the launching vehicle elements.

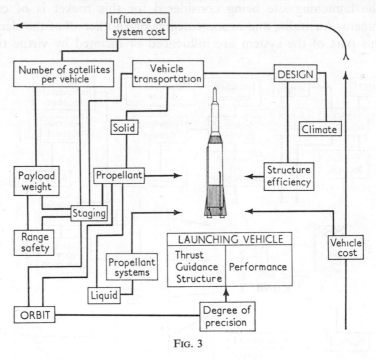

FIG. 3

3. *The Satellite* (*Fig. 4*)

Various sizes of satellites are being considered by different companies ranging from a few tens of pounds to a few hundreds of pounds in weight. I am discussing active satellites, which I believe have far greater potential than passive satellites, in spite of the inherent reliability of the latter. Being conservative for the moment and taking the weight of a good capacity active satellite as several hundreds of pounds, and also assuming that several satellites will not be launched by one vehicle then this tends to lead to a global system involving a pattern of satellites at

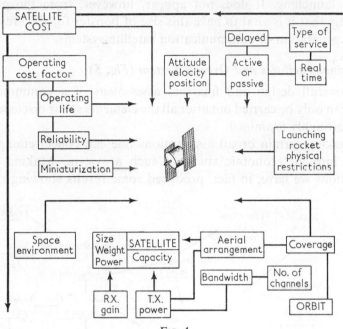

FIG. 4

medium altitudes (say between 4000 and 10,000 miles) each satellite maintained in its correct station, and not allowed a random position. To establish such a satellite in these conditions requires (with current techniques) a multi-stage launching vehicle with a lift-off thrust of at least 200,000 to 300,000 lb. BLUE STREAK lift-off thrust can be in the order of 300,000 lb and we therefore see that this rocket satisfies our requirement, whereas smaller rockets such as SCOUT, THOR, etc., certainly would not. If advantage is taken of developments of high energy upper stages for BLUE STREAK then payloads of significant size can be contemplated for the stationary 22,300 miles orbit, although the basic version with less exotic propellents in the upper stage might only put a few tens of pounds in this unique orbit.

We can therefore see that, with the choice of this BLUE STREAK booster and its launching facilities, there is some limitation on the size of satellite which can be launched but the important thing is that adequate payloads for a very comprehensive satellite communication system can be launched into satisfactory orbits.

Clearly the use of the larger booster such as ATLAS–AGENA B or ATLAS–CENTAUR will give a higher payload capability, including multiple payload launching. It does not appear, however, from the previous reasoning, that it is vital to have this size of booster for the creation of the first generation of communication satellite systems.

4. Terminal Facilities and Overall System (Fig. 5)

The overall design and financial assessment of a communication system can only be carried out after all the elements so far discussed have been thoroughly examined.

By making certain broad assumptions one can collect enough data to feed into an economic study of such a system. Making certain assumptions we have, in fact, produced some results showing the cost

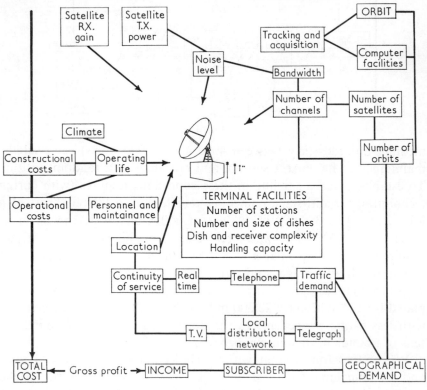

Fig. 5

of establishing and operating a system. These show that the cumulative revenue reaches the level of the cumulative expenditure within a few years of starting the system, even allowing for the large measure of research and development costs. Thereafter the gross profit, that is the difference between total revenue and total outlay, is of the order of tens of millions of pounds per annum. I would like to emphasize here, however, that the result mentioned is merely typical, falling within the bracket of results of such a study, and indeed one can only sensibly narrow the bracket by embarking on far more detailed consideration of all of the elements involved than has been done so far with this particular combination of equipment.

An analysis of expected communications traffic increase, is naturally the vital part of an economic study. The advent of new communications facilities via space will, I suggest, lead to a far greater build up of communication traffic than ever to be expected from orthodox trends, particularly in those parts of the world hitherto isolated and only recently starting to expand their own internal land telephone and television networks. However, even a very conservative approach to this traffic build up supports the order of operating returns that I have just mentioned.

Undoubtedly we shall see large differences between the new communications networks which will come into operation in the next few years since the relationship between the various factors involved are so complex and must inevitably lead to different solutions for different purposes. One thing is certain the new ability to communicate via space will be one of the most exciting and progressive events of our time.

SATELLITE COMMUNICATIONS SYSTEM PROVIDING CHANNEL DROPPING

E. K. SANDEMAN

British Aircraft Corporation, Ltd., Luton, Bedfordshire, England

I. INTRODUCTION

Channel dropping is the feature by virtue of which the satellite can provide communication simultaneously between a number of pairs of ground stations.

The origin of the paper was the construction of a model of a satellite communications system in order to make an assessment of the required r.f. power handling capacity in the satellite. It has been subsequently rather embellished in order to show that it is capable of being operated as a practical system and, in order to lend some substance to the original concept, by quoting illustrative numbers. A simple means is described for eliminating echo at the far end of 4-wire land lines associated with satellite radio links.

The results for the required r.f. power are presented as calculated when a particular p.c.m. code is used which reduces the loading of the satellite transmitter to a minimum. This code demands circuitry which is elaborate as compared with the circuitry of a p.c.m. coder based on what is effectively a digital voltmeter. For obvious reasons very many problems are ignored.

While the telecommunications system is described in terms of active satellites, it is equally suitable for a passive satellite system provided the ground transmitter power is made equal to 100 kW, provided the fact that no frequency shift takes place in the satellite is allowed for, and provided a suitable wavelength is used.

The telecommunications system described will also operate either with oriented 24-hr satellites or with any suitable number of unoriented satellites in random orbits provided the aerial and diplexing problem can be satisfactorily solved in the latter case.

Variations of the telecommunications system are described which will operate both for trunk services or area services. In the first case communication is provided by a number of circuits between 2 ground stations only, while in the area communications system each of a number of ground stations in a given communications area can communicate with each of the other ground stations in the same area. The satellite then in effect behaves like a telephone exchange: all that is necessary for 2 ground stations to get into communication is for each of the ground radio receiving stations to tune into one of the communication bands assigned to the radio transmitter of the other ground station. The feature afforded in this way is sometimes referred to as channel dropping.

II. DESCRIPTION OF SYSTEM

A. *Telecommunications System*

1. *Basic Requirements*

The communication system must meet the following requirements.

Channel Dropping

(*i*) It must permit a single satellite equipped with one or more transponders (receiver-transmitters) to provide a number m of conversations, as required, between any one of a number G of ground stations in the communications area of the satellite or satellite system, and any other ground station in the area. A total number of M such conversations must be capable of being carried on simultaneously. (The values of m and M are respectively determined by the traffic demand from the various ground stations and the traffic handling capacity of the satellite.) When $G = 2$, a trunk system results.

(*ii*) Communication between ground stations must be capable of being effected by processes confined to the relevant transmitting and receiving stations on the ground and by reference to a master control station issuing suitable frequency band allocations, either permanently or temporarily.

(*iii*) Cross talk between speech channels must be tolerable.

(*iv*) Speech distortion, whether due to response distortion, nonlinearity, or level quantizing (i.e. in p.c.m. systems) must be tolerable.

(*v*)   Adequate speech to noise ratio must be provided.

(*vi*) For any given number of channels the optimum compromise must be achieved between the requirements of minimum r.f. power in the satellite transmitter, minimum frequency bandwidth occupancy and simplicity of equipment (it is evidently going to be difficult to show that this condition has been met).

2. *System to Meet the Above Requirements*

The general arrangement of equipment at 2 ground stations J and W of a satellite communications system satisfying the main requirements is indicated in Fig. 1. These ground stations are considered to constitute 2 out of a number of ground stations which might be as large as 20. It will be seen that a ground station comprises a radio transmitter and receiver and all associated terminal gear and processing equipment.

The arrangement of Fig. 1 will in general apply regardless of the number of speech channels handled by the satellite and ground station, with the reservation that the detail interpretation of certain of the symbols in the diagram will probably, but not necessarily, vary with the number of channels handled at a ground station. The requirements for a ground station which will require to receive 1 channel from any of 20 other ground stations will obviously be different from those of a ground station handling 50 conversations from 1 other station only.

It is assumed that each ground station is permanently allocated a number of transmitting frequency bands designated by the letter T and a number, each such frequency band providing a transmitting channel carrying one telephone conversation from ground to the satellite. At the satellite each of the said frequency bands is transferred to another part of the spectrum where it is designated on Fig. 1 by the letter R and a number.

For instance, if transmission band T.470 is allocated to ground station J which requires to communicate with ground station W to which transmission band T.175 is allocated for the return transmission band, the subscriber at end J talks up to the satellite on band T.470, down from the satellite on R.470, while the subscriber at terminal W talks up to the satellite on band T.175 and down from the satellite on band R.175. The receiving system at terminal J must therefore tune into band R.175, while that at terminal W must tune into R.470.

It is assumed that at each terminal there is a terminal exchange not shown and that the single lines entering the picture from left and right

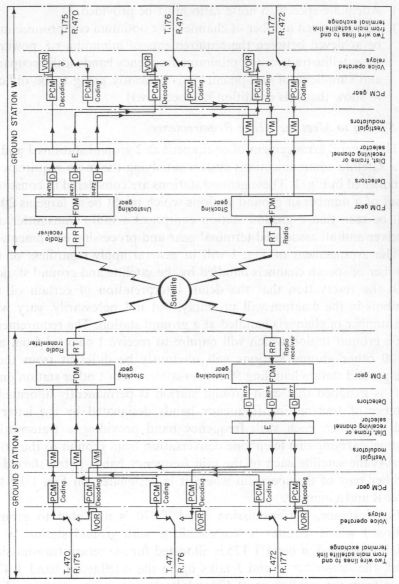

Fig. 1. Basic telecommunications layout.

represent lines from this exchange, which by virtue of the connexions made in the terminal exchanges become the 2-wire lines associated with the subscribers at each end. The fact that after passing through the exchange the circuits may pass through 4-wire land lines to the subscriber does not invalidate the discussion. In Fig. 1 the 2-wire lines at each terminal are labelled with the letters and numbers indicating the frequency bands used respectively:

(*a*) for carrying the subscriber's conversation up to the satellite; and

(*b*) for bringing the remote subscriber's reply down from the satellite.

It will be seen that in Fig. 1 circuits are set up for 3 simultaneous conversations between terminal J and terminal W. It may be imagined that these and other ground stations are all carrying on simultaneous conversations with one another via the satellite using other combinations of frequency bands.

All this is obvious, but it has to be said.

It may be remarked that all the R frequency bands must be separated from all the T frequency bands by a distance such that there is a spacing of some 100 to 300 Mc/s between the transmitting frequency band and the receiving frequency band which are nearest in the frequency domain.

Now consider the system of equipment by means of which a subscriber on 2-wire line T.470 (R.175) talks to the subscriber on 2-wire line T.175 (R.470). It is perhaps as well to emphasize at this stage that the lines designated as T.175 (R.470) etc. are 2-wire lines carrying ordinary conversations both ways. Whatever processes are introduced between the 2-wire line and the radio transmitter and the 2-wire line and the radio receiver, there nevertheless has to be some device inserted at the point where the 2-wire both-way line joins the 2-wire one-way line to the radio transmitter, and the 2-wire one-way line from the radio receiver, to prevent speech power arriving from the receiver from entering the line to the transmitter and so causing echo or singing.

a. Voice Operated Relay In Fig. 1 this device is represented by the block labelled v.o.r. indicating voice operated relay which, in the absence of speech voltages arriving from the radio receiver, is in such a position, that the 2-wire lines at each end of the circuit are connected to their respective radio transmitters. On the arrival of speech voltages from the radio receiver the relay is operated and the circuit from the radio receiver is connected to the 2-wire line. The circuit of the voice operated system is shown in Fig. 2(a), where a delay network is introduced to prevent clipping of the speech due to the operating time of the relay. The writer's experience on a link from Madrid to Buenos Aires some 30 years ago suggests, however, that such a delay network is not essential and inquiries from the G.P.O. have confirmed this view. On the Madrid–B.A. link, or even on the long-wave link from Rugby it would have been impossible to operate such a system, because of the poor speech to noise ratio on the radio receiver output. The alternative system in which the circuit is normally through from the 2-wire line to receiver, and speech voltages from the 2-wire line connect it to the radio trans-

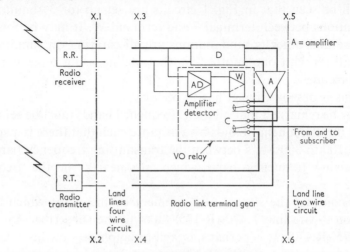

FIG. 2(a). Voice operated relays.

mitter, had to be used. With this system there is a possibility of both subscribers talking at once without knowing that the other is talking. This condition is known as lock-out and the probability of its occurrence increases with the transmission delay of the radio circuit. The system here proposed is made possible by the high signal to noise ratio due to p.c.m.

The voice operated relay would be designed to operate within 5 or 10 msec of the arrival of speech voltages from the receiver.

In the case of low satellites (not above 6000 or 8000 miles high), the hold-on time of the relay could be made the optimum to maintain continuous connection between syllables when the speech voltage falls. If any 4-wire land line, associated with one end of the satellite link has a propagation time greater than half the hold-on time any echo from the far end of the 4-wire line is returned to the speaker. The effect of such echoes for the time of propagation of satellite links with satellites not higher than 6000 to 8000 miles is understood to be tolerable, because the echo returns a comparatively short time after the speaker has stopped talking.

In the case of 24-hr satellites the echoes from the far end of associated

4-wire land lines are considered troublesome and for use with a 24-hr satellite circuit the hold-on time would have to be increased when associated 4-wire land lines have propagation times greater than a half the normal hold-on time. In this case the hold-on time would be made equal to the optimum for bridging intervals between syllables *or* twice the propagation time from any remote 4-wire terminal in any associated land line plus say 5 msec, whichever is the greater. Such a procedure will mean that the speakers at each end of the line will not be aware of the introduction of the increased hold-on time and will not hear any echo.

The order of magnitude of the maximum land line propagation times that it is required to deal with is believed to be about 100 msec. Experiment is required to determine whether with a zero propagation time between speaker and the v.o.r., a hold-on time of 200 msec is permissible. If so, the necessity for varying the hold-on time with length of circuit can be avoided.

The arrangement described avoids both lock out and echo, since a delay of 0·6 sec on a reply to a question is tolerable, this circuit effectively disposes of the bogy of delay in a 24-hr satellite link.

The last ditch defence of opponents of the 24-hr satellite is worth recording. They point out that, with any circuit which kills far end echo, it is impossible for the subscriber at either end to interrupt the talker while he is speaking. Without very special measures, this is of course true, but it is pertinent to point out that this argument applies equally to the low satellite case unless echo suppression is omitted. The sug gestion that echoes may be acceptable in the low satellite case is at variance with the statement by all proponents of the low satellite case, that the service provided by the satellite communications system must be as good as that provided by existing cable or land line circuits. To argue that the long delay militates against the 24-hr satellite, it is now necessary to demonstrate that the echo occurring in the case of the low satellite case is less objectionable than the loss of the privilege of being rude. Finally, a circuit to permit interruption exists.

b. Pulse Code Modulator Returning to Fig. 1 and following the circuit from the v.o.r. to the radio transmitter we arrive at a block marked p.c.m. which is a pulse code modulator. The speech voltage is sampled at intervals of 1/8000 sec during each of a series of time quanta of the same duration (i.e. 1/8000 sec) and during the next time quantum a pulse code of 7 time units or bits is sent out indicating the observed

C.S.–B

voltage level in magnitude and sign. This means that the duration of each of the code pulses is $1/(8000 \times 7) = 1/56,000$ sec.

This has transformed the voice frequency wave into what is virtually a telegraph signal: a series of time intervals during which a pulse, which is always of the same amplitude, is either present or not present. Such a system is sometimes called a binary code system.

c. Formation of Basic Channel Frequency Band Because of the fact that the information frequency band of the system of pulses at the output of the p.c.m. modulator goes down to zero frequency, it is not possible to extract a conventional single sideband for frequency transfer to suitable parts of the frequency domain, and the balance of balanced modulators is inadequate to give proper suppression of the unwanted sideband, so that these cannot be used for the same purpose. For this reason a preliminary band of frequencies is formed by modulating a suitable carrier frequency with the output of the p.c.m. coder and selecting, by means of filters and frequency response correctors, one sideband, a vestige of the other and the carrier frequency. The frequency response is so corrected that after detection the original modulating wave can be obtained in its original form: i.e. of a pulse train.

d. F.D.M. Stacking Equipment The frequency division multiplex stacking equipment is nothing more than a system of modulators and filters by virtue of which the basic channel frequency bands are transferred to the required parts of the frequency spectrum.

e. Trunk System In the case of a trunk system this is done by stacking 12 *basic channel frequency bands*, suitably transferred in frequency, side by side in the frequency domain to form a *basic group* of 12 channels. Five *basic group frequency bands*, suitably transferred in frequency, are stacked side by side in the frequency domain to form a *basic super group*. Any required number of *basic super groups* are transferred in frequency and stacked side by side. Illustrative numbers are given later and the resulting f.d.m. stacking system is called the basic f.d.m. system. The corresponding equipment is called the basic f.d.m. equipment. If one of the variations in the section "Variations" is used this will give rise to considerable, but systematic revision of the basic f.d.m. system.

f. Area System In the case of an area system, stacking of frequency bands may be based on the basic f.d.m. system, i.e. by using the minimum number of parts of the basic f.d.m. equipment to stack the comparatively few (as compared with the trunk system case) channels which are transmitted from each ground station.

Again if any of the variations proposed in section 3 are used, they will modify the system, but probably not so much as in the trunk case, since the requirements of flexibility will probably demand that quite a number of channels are individually p.c.m. coded as in the basic system on page 28.

At this stage it may be useful to remember that the frequency bands allocated to each transmitter are permanently allocated, and that these bands determine what parts of the f.d.m. system are put in or left out. Because the allocation of transmission bands is permanent, the set of f.d.m. stacking gear which stacks in the frequency domain the frequency bands carrying each channel is permanent; no tuning operation is associated with the f.d.m. stacking gear.

g. Radio Transmitter The output of the f.d.m. stacking gear is connected to the radio transmitter. In the radio transmitter the frequency band constituted by the stacked array of channel frequency bands is transformed in frequency to a suitable band of frequencies for radiation, and transmitted to the satellite. The preferred band of frequencies for a passive satellite system is in the neighbourhood of 6000 Mc/s and this figure has been quoted for an active satellite. The author finds it hard to understand why such a high frequency is proposed for this purpose, and this point is discussed later.

h. Satellite Communication Gear In the satellite the received frequency band is shifted by 100 to 300 Mc/s and retransmitted to ground. It is not proposed here to deal with the problems of reception and diplexing, which appear to be soluble. The power requirements in the satellite are considered below.

i. Main Ground Radio Receiver The main ground radio receiver is assumed to use a maser for initial r.f. amplification which, for an angle of elevation of the satellite of 10° and a corresponding angle of elevation of the receiving aerial array or dish, has an effective temperature of substantially 40°K, having regard to the noise radiation from the atmosphere, see Fig. 3(a) and (b), which are taken from Ref. 5.

The receiver is essentially a double or triple superhet with a reconstituted carrier frequency supplied at a suitable frequency. Since the incoming frequency band is subject to doppler shift, the required frequency of the reconstituted carrier is unknown and it must be possible to switch in a circuit providing automatic search in frequency and automatic frequency lock every time it is necessary to tune in.

The overall function of the main radio receiver is to select the incoming frequency band and transfer it in frequency to the frequency band occupied by the original information band at the output of the f.d.m. stacking gear at the transmitting station. This information band is then supplied to the f.d.m. unstacking equipment. This receiver is not to be confused with the multiple superhet receivers described below which accept the output of the main receiver as their input.

j. F.D.M. Unstacking Equipment and Receiving Channel Selector The function of this combination of equipment is to select and deliver to each v.o.r. 2-wire to 4-wire junction, the receiving channel from the remote ground station with which the transmitting side of the v.o.r. is connected because:

(*i*) the outgoing channel from the v.o.r. towards the f.d.m. stacking equipment is connected permanently to one channel frequency band;

(*ii*) the remote ground station is tuned into that channel frequency band.

k. Trunk System In general the arrangements at a trunk terminal will differ from those at a ground station, in an area communication system using channel dropping in one or two respects. At a trunk station a systematic arrangement of f.d.m. unstacking gear is essential and the demodulating frequencies will be *fixed*. Unstacking is effected by systematic sequential selection of super groups, groups and channels each of which is transferred to its basic position (see page 31) before selection in its basic filter. The outputs of the f.d.m. unstacking gear (i.e. each recovered basic channel frequency band) go to detectors D, the outputs of which are permanently wired (on a distribution frame represented by E on Fig. 1) to the v.o.r. junctions to the 2-wire circuits to the radio link terminal exchange.

1. Area System In this case the arrangements will not only depart from those at the trunk terminal, but will probably vary from ground station to ground station, according to whether particular v.o.r. junctions are permanently or temporarily associated with particular distant ground stations.

If all the v.o.r. junctions are permanently associated with particular remote ground stations so that no switching of receiving channels is required then, since the number of channels per ground station is only a small fraction of the total number of channels for which the basic f.d.m. system is designed, the f.d.m. system can be most conveniently realized by selecting from the basic f.d.m. gear the minimum combination of demodulators and filters which are sufficient to select the receiving channels permanently associated with the v.o.r.'s respectively. The selected channels, at the outputs of detectors D will be permanently wired to their associated channels as before.

If any or all v.o.r.'s are required to carry conversations to a number of remote ground stations, then a number of different arrangements are possible. All arrangements must be capable of connecting the receiving side of any v.o.r. to any of 1200 or 2000 receiving channels or whatever the total number of receiving channels is.

One proposed method replaces the f.d.m. unstacking gear, the detectors D and the frame, by the following equipment. *For each v.o.r. input circuit* what is in effect a *multiple detection superhet receiver* is constituted, using appropriate filters and modulators of the basic f.d.m. equipment in conjunction with one variable frequency band-pass filter and variable frequency oscillators.

The superhet receivers so constituted are not to be confused with the main radio receiver above. Their function is to tune into parts of the whole information band at the output of the main radio receiver and to select the individual channel frequency bands from it.

During the process of tuning, a variable tuning circuit makes a rough selection of the required supergroup (e.g. 3780 kc/s wide, see below), and a variable frequency oscillator in conjunction with a modulator transfers the selected supergroup to the basic supergroup frequency band where it is selected by a standard basic supergroup filter.

By means of a variable frequency oscillator and a modulator, the required group is transferred in frequency to the basic group frequency band where it is selected by a basic group filter.

Finally by means of another variable frequency oscillator and a modulator, the required channel is transferred in frequency to the basic channel frequency band where it is selected by a basic channel filter.

The output of the basic channel filter goes to a detector.

In practice the whole process could be very simply automated. The variable frequency oscillators would be variable in steps or would be replaced by a standard frequency generator emitting all the frequencies required which would be switched in by punching a keyboard. The tuning circuit selecting the required supergroup would also be provided either by a number of filters from the basic f.d.m. unstacking system, or else would be a variable band-pass tuning circuit adjustable in steps. Tuning of this circuit would also be effected by pressing keys.

The method of putting 2 ground stations into communication is then for each to state the transmission band it proposes to use, and each ground station then adjusts the f.d.m. gear or superhet receiver, associated with the v.o.r. connected to use the declared transmission band, so as to receive the receiving band associated with the transmission band from the other station.

3. Variations

a. Trunk System If a trunk system is used, 2 alternative systems present themselves:

(i) The standard G.P.O. or C.C.I.T. system of f.d.m. is used to stack voice frequency bands into groups of 12 channels occupying, say 48 kc/s and/or into supergroups of 60 channels. Any number of these groups or supergroups are supplied to a single p.c.m. coder in which they are sampled at a suitable rate and pulse coded in a suitable number of levels.

This system makes progressively more severe demands on circuit techniques as the bandwidth processed is increased and pulses become shorter, but for a bandwidth of reasonably modest dimensions, where circuit difficulties are not severe it has the advantage of replacing a number of p.c.m. coders by one. See Ref. 20 for B.T.L. experimental results on high speed p.c.m. coding.

The band of frequencies created as a result of the pulse coding will extend down to zero frequency and will have to be transferred to a suitable band (for frequency shift to higher frequencies) by a vestigial sideband process. The frequency bandwidth lost by the vestigial sideband filter attenuation slopes will be the same as the aggregate of the bandwidths lost due to the individual vestigial sideband systems in the basic system.

As compared with individual p.c.m. coding of speech channels using equal level quanta, this system is at some disadvantage because, since it is handling the sum of the voltages due to a number of channels, the number of level ranges has to be increased. This means that the number of time units in each character has to be increased so that a wider frequency bandwidth is required. As compared with p.c.m. coding of individual speech channels using optimumly tapered voltage level quanta this system is at a further disadvantage on frequency bandwidth since it is unable to take advantage of tapered level steps.

(*ii*) A block of n individual speech frequency channels is p.c.m. coded as follows. Individual p.c.m. coding of each speech channel is effected, and pulses of $1/n$ the duration of the time unit appropriate to one channel are used for each channel. The pulses of each channel are then stacked contiguously in time so that the information pulses carrying the code characters are interleaved in time, so providing again a system of time division multiplex.

This system has the disadvantage of individual p.c.m. coding, but the advantage of a rather narrower frequency bandwidth than system (*i*) immediately above, even if tapered level quanta are not used: with a power economy code tapered level quanta will probably not be used.

b. Application of Above Variations to Channel Dropping Both the above systems can be made applicable to channel dropping provided the whole block of channels constituting one system of time division multiplex is transmitted from one ground station only and received only at one other ground station, the corresponding frequency band being transferred to appropriate positions in the frequency domain by a suitable system of f.d.m.

The resulting f.d.m. system will lack the systematic regularity of the basic system, conceived above, for instance in the case of a trunk system, and more thinking about the layout of frequency bands and modulating frequencies etc. will be required. In principle, however, no serious technical difficulty appears.

From an operating point of view it appears that any block of channels processed in the same way will have to be allocated permanently to one pair of ground stations to carry their mutual traffic. This may be no disadvantage even in an area system, and is certainly not one in a trunk system.

c. Allowance for Doppler Shift This is negligibly small in the case of a 24-hr satellite system, whether used as a trunk system or as an area system.

Range of Doppler Shift The magnitude of the doppler shift from satellite to ground or vice versa depends on:

(*a*) the wavelength

(*b*) the line of sight velocity of satellite with respect to ground station.

The latter has 2 components contributed respectively by:

(*i*) velocity of satellite in the inertial frame and

(*ii*) velocity of ground station in inertial frame due to rotation of the earth.

The ground stations for which the satellite's contribution to line of sight velocity are a maximum are those on the equator at the edge of the coverage area, i.e. at the lowest angle of elevation of the satellite which is permissible.

Earth's Contribution to Line of Sight Velocity The earth's peripheral velocity at the equator is 464 m sec^{-1} and the contribution to line of sight velocity is therefore 464 cos ψ m sec^{-1}, where $\psi =$ the angle of elevation of the satellite above the horizon.

If the lower limiting angle of elevation of the satellite is 10°, the earth's maximum contribution to line of sight velocity at the equator for a satellite in an equatorial orbit is:

$$v_e = 464 \times 0{\cdot}9848 = 457 \text{ m sec}^{-1}.$$

Since we are looking for maximum variations of satellite velocity due to the earth's component of velocity adding or subtracting to or from that satellite, the ground stations are always taken as being at the edges of the coverage area where $\psi = 10°$, so that the earth's contribution is constant at $\pm$ 457 m sec^{-1}.

Table I below gives the extreme limits of variation of the doppler frequency due to both causes for different satellite heights and 2 wavelengths:

TABLE I. Satellite extreme maximum doppler shifts for different wavelengths
Equatorial orbits
Minimum angle of elevation of satellite $= 10°$
Sense of satellite rotation the same as earth's
Doppler shifts are from ground to satellite or vice versa

Satellite height (miles)	Satellite orbital velocity (km sec^{-1})	Satellite component of limiting doppler frequencies		
		Sight line velocity (km sec^{-1})	$\lambda = 5$ cm (kc/s)	$\lambda = 15$ cm (kc/s)
3,000	5·97	3·347	57·7	19·2
6,000	5·0	1·958	30·1	10·0
8,000	4·56	1·487	20·6	6·9
12,000	3·496	0·854	8·0	2·7
18,000	3·364	0·6	1·85	0·62
22,300 (stationary)	3·077	0·457	zero	zero

Maximum Relative Shift on Channels from Different Ground Stations
Case 1. Shift at Satellite The maximum relative shift occurs when one channel is shifted up in frequency by the doppler shift from ground to satellite and the other is shifted down by the same amount. The shift is therefore twice the values of shift given in Table I.

Case 2. Shift at Ground The maximum shift which can occur here between 2 channels received on the ground is when one ground station is receiving from a second on which the shifts add and from a third on which the shifts cancel. The net result is again that the maximum relative shift is twice the value of shift in Table I.

Remedy The band of frequency allocated to any ground station must be separated from frequency bands nearest in the frequency domain allocated to another ground station or ground stations by at least twice the relative doppler shift which can occur between transmissions from that ground station and the said other ground stations.

B. *Illustrative Numbers and Magnitudes in the Basic System*

The word basic here is intended to limit the description to a trunk system transmitting a required number of supergroups of individually p.c.m. coded channels. This permits a systematic presentation of the way in which channels are stacked in the frequency domain.

The basic system serves as a point of departure for describing, for instance, the equipment used at a ground station which is part of an area communications system. For this purpose suitable modifications must be made as described above by leaving out equipment or by introducing special multiple detection superhet receivers in place of the f.d.m. gear and distributing frame etc.

Speech Range The system as described transmits a speech range from 300 to 3400 c/s. This range has been chosen because it is the standard speech range at present in use in G.P.O. f.d.m. systems for transmission over coaxial cable. A minimum of processing at terminals is therefore required. It will also handle the speech range from 200–3050 c/s used on the Transatlantic Cable TAT. 1 from Newfoundland to Oban.

1. *Pulse Code Modulation*

a. Pulse Coding of Individual Speech Channels Each incoming speech channel will be assumed to have no frequency components higher than 4 kc/s. If in any case there are such components, a filter must be inserted to remove them.

The incoming speech voltages are first passed through a constant volume amplifier to take out level variations due to speakers and tolerances in equipment. The resulting band of frequencies is then pulse coded in a tapered level step coding device of optimized taper so as to reduce quantizing noise to a minimum. In general this means that the level steps increase in size with increase of speech voltage level (deviation from zero whether positive or negative). The writer is not aware of any work which has been done to optimize the tapering. Tapering may be carried out by any suitable nonlinear system: either a nonlinear transducer followed by a linear level quantizer, or by the use of a nonlinear level quantizer. With power economy code, taper may be omitted, see below.

b. Pulse Code A pulse code is characterized by a number of different factors among which are the following. The method of quantizing the voltage level into level quanta or steps including the number of steps into which the voltage conveying the information is divided. The number of voltage levels used for the pulses providing the code. The number of time units or bits in the code. The way in which the code characters are associated with the voltage level quanta they represent.

Method of Quantization In general voltage levels above some level which has a sufficiently small probability of being exceeded are neglected and it will be assumed that a limiter is inserted to remove them. A suitable setting for the limiter is such that it excludes all levels which have a probability less than 1 % of being exceeded. The limiter will be set to exclude both positive and negative levels outside this limit.

The quantization of the positive level range will now be considered, it being understood that the quantization of the negative level range is identical.

The positive level range may be divided either into equal level steps, or into steps such that the level quanta they include have an equal probability of occurrence, which means that the voltage level steps are tapered, getting larger as the voltage level is increased, or some other system of tapered level steps may be used.

Through the courtesy of Messrs R. O. Carter, H. B. Law and W. H. White of the G.P.O. Research Station at Dollis Hill, London, the writer has been given a demonstration of p.c.m. using both equal and tapered level steps, and although the taper was not optimized, the demonstration left no doubt in his mind that either a 7 unit code with equal level steps, or a 6 unit code with tapered level steps will provide speech of acceptable quality and intelligibility.

Binary Code It is proposed to use a code in which pulses at only 2 levels are used: of zero or finite amplitude: all finite pulses being of the same voltage amplitude. This is called a binary code.

Number of Time Units or Bits in the Code Three codes are considered: An equal level step code using 7 bits. An equal probability level step code using 6 bits. A conventional binary digital equal level step code, see Appendix II.

In each case, for the purpose of calculating signal to noise ratio it is assumed that an extra time unit is used to provide time for essential processes during coding and decoding.

Association of Code Characters with the Level Quanta they Represent In the case of the equal probability 6 bit p.c.m. system just described, since each level range has an equal probability of occurrence, and since with a given number of bits in a binary code, there are a limited number of characters, all of which are normally used to indicate levels, the mean pulse rate and so the probability of a pulse occurring in each time unit is independent of the way in which the code characters are associated with voltage level quanta.

Seven Unit Power Economy Code In the case of the equal level step p.c.m. system, just described, by arranging that as far as possible, the higher the level at which a level quantum occurs, the more pulses occur in the code, and the lower the level the fewer pulses occur in the code, the mean pulse rate is minimized and so that probability of a pulse occurring in any time unit is reduced.

Example An example of a 7 bit binary code is shown in Table II in which the last named conditions are substantially realized. In Table II positive voltage level quanta are distinguished by different codes from negative level quanta occurring at the same scalar value of voltage. Each level range is represented by a number indicating the number of voltage quanta from zero voltage to its upper (most positive or most negative) limit; and a plus or minus sign to indicate whether the level corresponds to a positive or a negative voltage. An exception is the level range centred on zero voltage which is represented by both its limiting levels: $-\frac{1}{2}$ and $+\frac{1}{2}$.

While this code leads to the minimum power demand on the satellite transmitter, it does not lend itself to simple or conventional means of coding and decoding, for instance by means of digital voltmeters and

TABLE II. 7 Bit power economy code

Level range (defined by upper scalar limit)	No. of pulses per code character	No. of levels represented by character of given number of pulses
$\pm \frac{1}{2}$	0	1
$+ 1\frac{1}{2}$ to $+ 4\frac{1}{2}$ and $- 1\frac{1}{2}$ to $- 3\frac{1}{2}$	1	7
$+ 5\frac{1}{2}$ to $+ 14\frac{1}{2}$ and $- 4\frac{1}{2}$ to $- 14\frac{1}{2}$	2	21
$+ 15\frac{1}{2}$ to $+ 32\frac{1}{2}$ and $- 15\frac{1}{2}$ to $- 31\frac{1}{2}$	3	35
$+ 33\frac{1}{2}$ to $+ 49\frac{1}{2}$ and $- 32\frac{1}{2}$ to $- 49\frac{1}{2}$	4	35
$+ 50\frac{1}{2}$ to $+ 60\frac{1}{2}$ and $- 50\frac{1}{2}$ to $- 59\frac{1}{2}$	5	21
$+ 61\frac{1}{2}$ to $+ 63\frac{1}{2}$ and $- 60\frac{1}{2}$ to $- 63\frac{1}{2}$	6	7
Total number of levels and codes		127

conventional digital to analogue converters. This code leads to a required satellite r.f. power which is a fifth of that required for a conventional p.c.m. code, see Appendix II.

A proposed coding and decoding means for such a code is described in Ref. 16.

c. Essential Information Band Occupied by Individual P.C.M. Channel = Nyquist Band With a speech frequence range from 300 to 3400 c/s it is possible to insert filters which will ensure that no information frequencies occur higher than 4 kc/s. It is therefore proposed to sample each individual speech wave at 8000 c/s corresponding to a time quantum of 1/8000 sec. In addition to the 7 units in the characters defining level ranges, with the type of code illustrated in Fig. 2, it may be necessary to provide another time unit for switching purposes, so that in effect an 8-unit code is required. For simplicity in presentation, however, 7 time units are assumed. With a sampling rate of 8 kc/s, this gives 56,000 time units per sec, contained in an essential information band from 0 to 28 kc/s.

Pulse Coding of a Block of f.d.m. Channels In this case since the number of p.c.m. coders and decoders is reduced, the comparatively complicated equipment required for the use of the type of code indicated in Table II is more easily justified.

2. F.D.M. Stacking of Channels (see Fig. 2(b))

With the exception of the first stage of modulation, it is not proposed to give in detail the modulation frequencies used for frequency transfer of the basic p.c.m. channel into its position in channel groups, for transfer of basic groups into position in supergroups etc. The proposed system does not stand or fall by virtue of the numerical values given below, although evidently without some such indication the proposals might appear rather nebulous. It is, however, felt that the numerical values chosen are very near to values which might be used in a practical system.

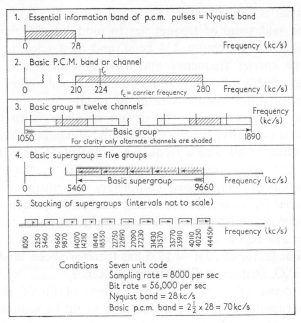

FIG. 2(b). F.D.M. stacking (Illustrative example for specified conditions).

The numbers given below are obtained by taking the positions in the frequency domain of G.P.O. or C.C.I.T. standard groups and basic groups, supergroups and basic supergroups, and multiplying them by the ratio of the p.c.m. bandwidth derived in the next section to the bandwidth of 4 kc/s used in the G.P.O. or C.C.I.T. standard f.d.m. stacking system. By this means it is hoped that no major blunders will have been committed in making demands on filters or modulation processes which cannot be met. See Refs. 8, 9, and 10.

a. Formation of Basic P.C.M. Channel from 210 to 280 kc/s: 70 kc/s wide

This channel corresponds to the basic 4 kc/s channel in the G.P.O. or C.C.I.T. stacking system, but the equipment for deriving it has no corresponding equivalent in the G.P.O. 4 kc/s system. The nearest corresponding process occurs in the derivation of the channels of a sub group in the G.P.O. economy system of frequency stacking using a speech frequency range of 200 to 3050 c/s (see Ref. 3) and this does not use a vestigial sideband.

Since the essential information frequencies at the output of each p.c.m. coder occupy a range *from zero* to 28 kc/s, it is necessary to constitute what is here called a *basic p.c.m. channel*, which is obtained by using the said band of frequencies to modulate a suitable carrier frequency (suggested location 224 kc/s), and then selecting the upper sideband, the carrier frequency and a suitable vestige of the lower sideband, by means of filters and equalizers so designed that after simple detection a flat frequency response results.

Experience suggests that if the design of filters is to be realized without undue difficulty, the upper sideband (including filter attenuation band) must be allowed a frequency range of 56 kc/s (i.e. so that the rise of filter attenuation is not too sharp) and the lower sideband may be conveniently suppressed completely at a distance of 14 kc/s from the carrier frequency. The resulting band of frequencies which has to be considered for frequency stacking purposes then extends from $224 - 14 = 210$ kc/s to $224 + 56 = 280$ kc/s and so has a bandwidth of 70 kc/s. The factor of $70/4 = 17·5$ has therefore been used in scaling from the G.P.O. frequency dispositions in Ref. 7.

At the expense of more expensive designs of filters, the effective frequency occupancy of the upper sideband might be reduced.

It is impossible at this stage to compare the respective demands of economy in money and economy in frequency bandwidth.

It will be understood that the output of each p.c.m. coder on each individual speech channel in the basic system, is used to modulate a carrier of 224 kc/s and so to produce, as described above, a band of information frequencies extending from 210 kc/s to 280 kc/s which constitutes what is here called the basic p.c.m. channel.

b. Trunk System Case

Formation of Basic F.D.M. Group of 12 P.C.M. Channels from 1050 to

1890 kc/s: 840 kc/s wide By intermodulating each of 12 basic p.c.m. channel frequency bands respectively with 12 suitable frequencies and selecting the lower sideband, the 12 p.c.m. channels are stacked contiguously in the frequency domain so as to occupy the frequency range from 1050 to 1890 kc/s and so constitute a basic group.

Formation of Basic Supergroup of 60 P.C.M. Channels from 5460 to 9660 kc/s: 4200 kc/s wide By intermodulating each of 5 basic groups of p.c.m. channels respectively with 5 suitable frequencies and selecting the lower sidebands, the 5 groups are stacked contiguously in the frequency domain so as to occupy the frequency range from 5460 to 9660 kc/s and so constitute a supergroup.

Stacking of Supergroups in the Frequency Domain By intermodulating each basic supergroup with a suitable different frequency and selecting the lower sideband, supergroups may be stacked in the frequency domain so as to occupy frequency bands as indicated below for the first 10 supergroups. The frequency spacings between the first 3 supergroups are 210 kc/s and between all other groups are 140 kc/s.

TABLE III. First 10 supergroups

Supergroup No.	Frequency range (kc/s)	Supergroup No.	Frequency range (kc/s)
1	1050–5250	6	22,890–27,090
2	5460–9660	7	27,230–31,430
3	9870–14,070	8	31,570–35,770
4	14,210–18,410	9	35,910–40,110
5	18,550–22,750	10	40,250–44,450

Ten supergroups containing 600 channels therefore occupy a frequency range of substantially 44 Mc/s. It is perhaps worth noting in passing that 44 Mc/s is rather a wide band even for a solid state maser, according to M. Brotherton of the Bell Laboratories who gives a figure for the amplification band of a solid state maser as 25 Mc/s, see Ref. 4. While this may pose a problem in the case of a trunk system in which as many as 600 channels terminate at one ground station, it does not raise any such problems in the case of area types of communication systems in which the number of channels dropped at any one ground station is unlikely to exceed or even to reach 300 channels, for which the required bandwidth is just over 22 Mc/s.

Ref. 25 states that Bell Telephone Laboratories are designing masers to provide the bandwidth needed for communication satellites.

F.D.M. Stacking in Area Case The deviations from the trunk case have already been sufficiently indicated in section 2(*f*).

C. *Radiated Information Band*

Trunk or Basic System The information band created by the f.d.m. stacking process extends substantially from 1·050 Mc/s to 44·45 Mc/s, and it is required to transfer this band 43·4 Mc/s wide to the neighbourhood of some frequency in the band from 1 to 10,000 Mc/s. In the present discussion, 2 frequencies will be discussed 2000 Mc/s and 6000 Mc/s and it will be assumed that by a suitable process of modulation (possibly double modulation) the frequency band from 1·05 to 43·4 Mc/s is transferred to one of the bands:

$$5956·6–6000 \text{ Mc/s or } 1956·6–2000 \text{ Mc/s}$$

The resulting band is then amplified and radiated. The alternative of a final stage of modulation instead of a final stage of amplification is not excluded.

Area System It is hardly necessary to explain that each ground station in an area system effectively uses such parts of the basic f.d.m. stacking gear that cause it to radiate on its allocated channel frequency bands.

D. *Adjustment of R.F. Power Radiated from Ground Transmitters and Satellite*

Provision has to be made so that whether the system is a trunk or an area communication system, during the period of transmission of a pulse on any one channel, the power loading of the satellite by that channel is of specified magnitude, regardless of the location of the ground transmitter from which the pulse originates. The specified magnitude is of course the same for all ground stations and is such that the design value of statistical loading of the transmitter by all channels occurs.

E. *Choice of Wavelength or Radio Frequency**

1. *Passive Satellite*

In the case of a passive satellite it can be shown that with 100 kW on the ground, a 100-ft diameter dish on the ground, and a wavelength of 5 cm corresponding to a frequency of 6000 Mc/s, the power re-radiated

---

*For U.S. activities in relation to choice of wavelength for satellite communications see Ref. 18.

from a spherical satellite of 140 ft diameter at a distance of 5000 miles (the extreme distance for a satellite height of 3000 miles), is 1 W. This compares unfavourably with the power required to be radiated from an active satellite at a height of 3000 miles for 600 two-way channels and a zero gain all round looking aerial system in the satellite, which is 6 W, i.e. corresponding to 12 W for gain = 0·5.

With a longer wavelength the aerial gain of the ground dish would be reduced and the power radiated from the satellite reduced in proportion.

The choice of a wavelength of 5 cm for the passive satellite which has been announced by the A.T. & T. and Bell System therefore appears to be one with which it is difficult to quarrel on the grounds that the wavelength is too short. The fact that the powers in Table IV are pessimistic, see conclusions, is unlikely to lead later to a very much smaller power requirement.

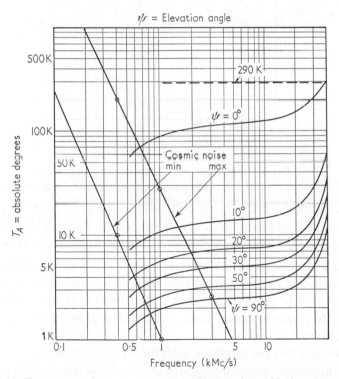

FIG. 3(a). Temperature due to oxygen absorption seen by an ideal antenna as a function of elevation angle ψ and frequency.

(*By courtesy of J. R. Pierce and the American I.A.S.*)

2. Active Satellite

The range of wavelengths from 30 cm (1 kMc/s) to 3 cm (10 kMc/s) has been proposed for satellite communications. Figures 3(a) and 4(a) show why.

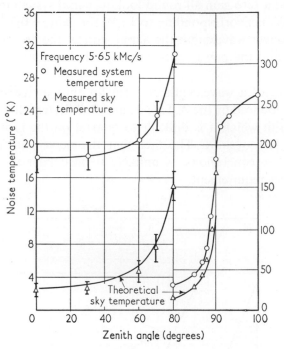

FIG. 3(b). Measured and calculated sky temperature.

(*By courtesy of J. R. Pierce and the American I.A.S.*)

The writer considers that a wavelength in the neighbourhood of 15 cm (2 kMc/s) should be considered for the following reasons.

(1) At 2 kMc/s, see Fig. 3(a) cosmic or galactic noise has fallen below atmospheric noise for most probable minimum angles of elevation (5° to 10°) of received ray above the receiving station, while above 2 kMc/s atmospheric noise increases slightly with increase of frequency.

(2) Noise due to rain increases with increase of frequency.

(3) The apparent earth temperature as seen from the satellite decreases with decrease of frequency.

(4) Above 2 Mc/s tropospheric attenuation (Fig. 4(a)) which is small

and attenuation due to rain, which can be quite large (see below), increase with increase of frequency.

(5) From 2 kMc/s to 4 kMc/s is the range in which it will probably be most easy to realize thermionic valves of adequate r.f. power handling capacity for use in ground transmitters.

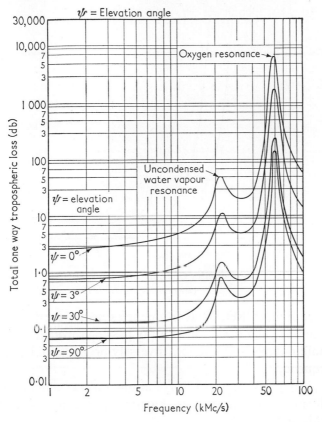

FIG. 4(a). Total one way tropospheric absorption.
(*By courtesy of H. J. Pratt and the American I.R.E.*)

(6) The half power points of the receiving polar diagram of the 100 ft diameter receiving dish which is proposed, are at ± 18 min of arc from the axis of the aerial for a wavelength of 15 cm corresponding to 2 kMc/s. This means that the direction of the dishes must be directed to this accuracy. As the frequency is increased tolerance is reduced. For instance at 6000 kMc/s the tolerance is ± 3 min of

arc which may be difficult to meet on satellites other than 24-hr satellites.

a. Ionospheric Effects According to Ref. 12 the attenuation of the ionosphere, to a ray passing right through it, is 0·3 db for a frequency of 0·1 kMc/s and decreases with increase of frequency.

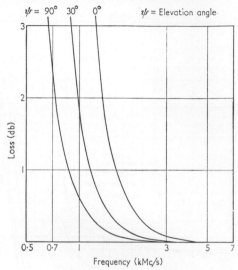

$\psi = 90°\ 30°\ 0°$ ψ = Elevation angle

Loss (db)

Frequency (kMc/s)

FIG. 4(b). One way Faraday loss.
(*By courtesy of H. J. Pratt and the American I.R.E.*)

Also according to Ref. 12 at frequencies near the critical frequency the maximum angular deviation of a ray passing through the ionosphere is a few milliradians and falls off inversely proportional to the square of the frequency.

Figure 4(b) gives the value of Faraday rotation loss for linearly polarized waves. Thus it appears that ionosphere effects are negligible above 2 kMc/s.

b. Size of Aerial in the Satellite The power gain required for a dish in the satellite in order that the 3 db points of the polar diagram should intersect the earth in a circular locus from which the angle of elevation of the satellite is 10°, is 28·8 assuming 50% efficiency for the aerial. On this basis the required diameter of dish is 70 cm, which appears to be acceptable.

3. *Availability of Radio Frequency Bands*

The following extract from Ref. 11 is given to show that the whole of the radio frequency range from 1 to 6 kMc/s is already allocated. The range above 8·4 kMc/s has been omitted, since the writer does not consider that this range is important.

In the summary of allocations below:

Region 1 = Africa, Europe and U.S.S.R.
Region 2 = North and South America.
Region 3 = Asia, South East Asia and Australasia.

TABLE IV. Recommended radio frequency allocations in band from 1000 to 10,000 Mc/s

Frequencies in Mc/s

Region 1	Region 2	Region 3
960–1215 Mc/s	AERONAUTICAL RADIONAVIGATION	
1215–1300	RADIOLOCATION Amateur	
1300–1350	AERONAUTICAL RADIONAVIGATION Radiolocation	
1350–1400 FIXED MOBILE RADIOLOCATION	1350–1400	RADIOLOCATION
1400–1427	RADIO ASTRONOMY	
1427–1429	SPACE FIXED MOBILE except aeronautical mobile EARTH-SPACE	
1429–1535 FIXED MOBILE except aeronautical mobile	1429–1435 FIXED MOBILE ———————— 1435–1535 MOBILE Fixed	1429–1535 FIXED MOBILE
1535–1660	AERONAUTICAL RADIONAVIGATION	
1660–1700	METEOROLOGICAL AIDS FIXED MOBILE except aeronautical mobile	
1700–1710 FIXED Space Mobile Earth-Space	1700–1710	FIXED MOBILE Space Earth-Space

Region 1	Region 2	Region 3
1710–2290 FIXED Mobile	1710–2290	 FIXED Mobile
2290–2300 FIXED Space Mobile Earth-Space	2290–2300	 FIXED MOBILE Space Earth-Space
2300–2450 FIXED Amateur Mobile Radiolocation	2300–2450	 RADIOLOCATION Amateur Fixed Mobile
2450–2550 FIXED MOBILE Radiolocation	2450–2550	 FIXED MOBILE RADIOLOCATION
2550–2700	FIXED MOBILE	
2700–2900	AERONAUTICAL RADIONAVIGATION Radiolocation	
2900–3100	RADIONAVIGATION Radiolocation	
3100–3300	RADIOLOCATION	
3300–3400 RADIOLOCATION	3300–3500	 RADIOLOCATION Amateur
3400–3600 FIXED MOBILE Radiolocation	3500–3700 FIXED MOBILE RADIOLOCATION	3500–3700 RADIOLOCATION Fixed Mobile
3600–4200 FIXED Mobile	3700–4200	 FIXED MOBILE
4200–4400	AERONAUTICAL RADIONAVIGATION	

Region 1	Region 2	Region 3
4400–5000	FIXED MOBILE	
5000–5250	AERONAUTICAL RADIONAVIGATION	
5250–5255	RADIOLOCATION Space Earth-Space	
5255–5350	RADIOLOCATION	
5350–5460	AERONAUTICAL RADIONAVIGATION Radiolocation	
5460–5470	RADIONAVIGATION Radiolocation	
5470–5650	MARITIME RADIONAVIGATION Radiolocation	
5650–5850	RADIOLOCATION Amateur	
5850–5925 FIXED MOBILE	5850–5925 RADIOLOCATION Amateur	5850–5925 FIXED MOBILE Radiolocation

The highly occupied condition of the frequency bands suitable for satellite communications suggests a requirement for sharing frequency bands with other services. In view of the extremely low levels received from the satellite there is a considerable probability that such sharing will be unsatisfactory. This points to the need for experimental work to explore the situation.

4. Effect of Rain

From Fig. 7 of Ref. 2, assuming that it is permissible to extrapolate from 10 cm as far as 15 cm and inserting the correction for 0°C to give a maximum attenuation at the longer wavelengths, the attenuation due to rain may be shown as having typical values as in Table V.

TABLE V. Attenuation over a 10-mile path (db)

	Great Britain			East Indies
Rainfall mm hr^{-1}	$p = 13$	$p = 26$	$p = 52$	300
Times per year that 3 min rainfall is exceeded	12	4	1	1
6 kMc/s	1·0 db	2·0 db	4·0 db	23·1 db
3 kMc/s	0·12 db	0·24 db	0·48 db	2·9 db
2 kMc/s	0·006 db	0·012 db	0·024 db	0·13 db

This does not provide an overwhelming argument in favour of 15 cm as opposed to 5 cm, for instance in Great Britain, and the information for the rest of the world is inadequate. None the less it is the opinion of the writer that, if frequency bandwidth is available, the wavelength should be as long as the size of satellite aerial permits, provided it is not longer than 15 cm.

III. REQUIRED R.F. PEAK POWER OF SATELLITE TRANSMITTER

A. Satellite–Earth Geometry

The inset diagram in Figs. 5, 6, and 7 shows the earth of radius r, the satellite at S at height h above the surface of the earth, and two ground stations G_1 and G_2 at the limit of satellite communication as determined by the minimum permissible value of ψ the angle of elevation of the satellite above the earth's horizon at the position of the ground station. The line G_1H is the horizontal at the ground station G_1

$\eta = OSG_1 =$ the angle subtended at the satellite by the ground station G_1 and the centre of the earth.

$\psi =$ the minimum permissible angle of elevation of the satellite above a ground station horizon for satisfactory communication.

$SG_1 = R_{max} =$ maximum range from satellite to ground station for satisfactory communication.

Figure 5 shows plots of η and the length of great circle arc G_1G_2 plotted against satellite height.

Figure 6 shows a plot of maximum range from ground station to satellite and also surface diameter of satellite coverage area plotted against satellite height for $\psi = 10°$.

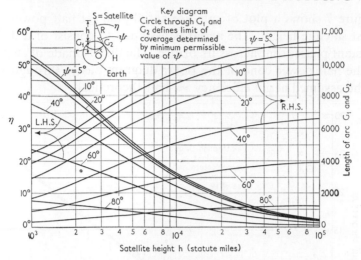

FIG. 5. Satellite–Earth geometry.

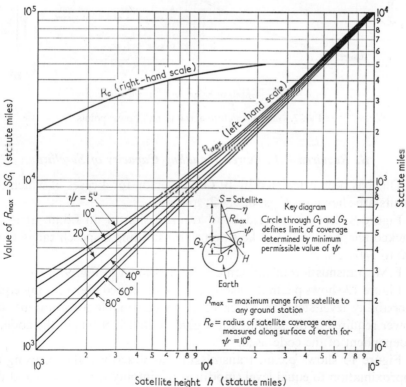

FIG. 6. Satellite–Earth geometry.

Figure 7 shows a plot of satellite aerial gain (at half power points) against satellite height. The aerial is assumed to consist of a dish with a constant phase front across its orifice. The dish is made of such a size that the half power cone of the polar diagram passes through G_1 and G_2. The assumed aerial efficiency = half theoretical.

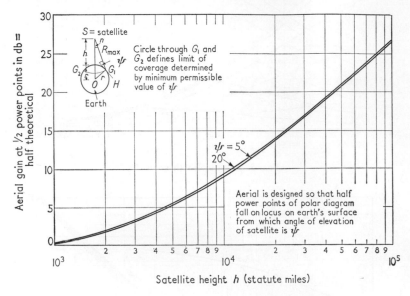

FIG. 7. Gain of satellite aerial at half power points.

B. Required R.F. Power Handling Capacity of Satellite

This has been calculated by B. G. Anderson for 4 cases of which the results are shown on Figs. 8, 9, 12, and 13.

Figure 8 shows single sideband transmission of an f.d.m. array of stacked telephone bands 4 kc/s apart transmitting a speech range from 300 to 3400 c/s.

F.M. transmission of the same array is shown in Fig. 9.

Figure 12 shows p.c.m. transmission, as described above, using equal probability level steps, in which case, on reasonable assumptions, the power required in the satellite for a given number of bits in the code, is independent of the code used.

Figure 13 shows p.c.m. transmission, as described above, using an approximation to equal level (modified to simplify calculation) and the 7 bit code, described above, designed for minimum satellite power.

1. Basis of Computation

The following conditions are common to Figs. 8, 9, 12, and 13.

T.A.S.I. (time assignment speech interpolation) is not used, see Ref. 6. The noise accompanying the speech at the input to the satellite link (e.g. at the outputs of the 2-wire circuits in Fig. 1) is neglected.

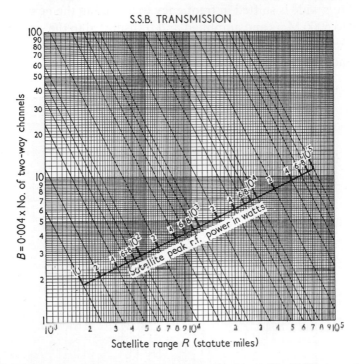

Fɪɢ. 8. Required peak r.f. power of single transponder in satellite (S.S.B. transmission). 1. Short average speech to noise ratio = 37 db for ψ = 5°: 2. Effective maser temperature = 39°K: 3. Diameter of ground receiver dish = 100 ft: 4. Satellite aerial gain 0·5: 5. Bandwidth of speech channel = 3400–300 c/s.

Satellite aerial power gain = 0·5. Receiver ground dish diameter = 100 ft. Receiver ground dish efficiency = 50% of theoretical. Maser effective temperature = 39°K corresponding to 5° elevation of satellite (Ref. 5). Speech range transmitted = 300 to 3400 c/s. The distribution of speech levels and the ratios of peak to mean power for different numbers of channels are taken from figures in Ref. 1, e.g. see Fig. 10.

a. S.S.B. Transmission and F.M. Transmission (Figs. 8 and 9) The short average mean speech power to mean noise power ratio on stations at edge of coverage area, i.e. where angle of elevation of satellite = 5°, is made equal to 37 db.

b. S.S.B. Transmission (Fig. 8) In the s.s.b. system, the band of frequencies at the output of a conventional G.P.O. f.d.m. stacking system containing channels spaced at 4 kc/s and so occupying a frequency band 2 or 3 Mc/s wide is shifted in frequency to a band in the neighbourhood of 2000 to 6000 Mc/s and amplified and transmitted to the satellite where it is shifted in frequency by between 100 to 300 Mc/s and retransmitted to ground.

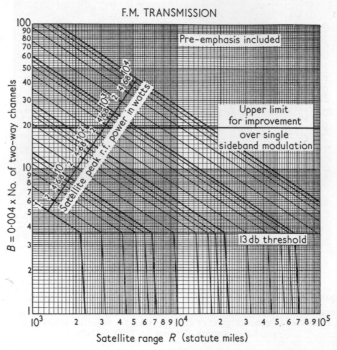

FIG. 9. Required peak r.f. power of single transponder in satellite (F.M. transmission)
1. Short average speech to noise ratio = 37 db for 5° elevation of satellite: 2. Max. deviation Δ = 20 Mc/s: 3. Effective master temperature = 39°K: 4. Diameter of ground receiver dish = 100 ft: 5. Satellite aerial gain 0·5: 6. Bandwidth of speech channel = 3400–300 c/s.

c. F.M. Transmission (Fig. 9)
$$\text{Peak frequency swing} = \pm 20 \text{ Mc/s}$$

In the f.m. system, the band of frequencies of some Mc/s wide at the output of the f.d.m. stacking gear is used to modulate the frequency of a carrier wave in the neighbourhood of 2000 to 6000 Mc/s, with a peak amplitude of swing of $\pm$ 20 Mc/s. The resulting wave is transmitted to the satellite where it is amplified, shifted in frequency by between 100 to 300 Mc/s and retransmitted to ground.

On account of its technical difficulties, the use of frequency feedback, as described for instance in Ref. 7, is not used. This means that the requirement that the signal to noise ratio at the input to the limiter in the f.m. receiver on the ground shall be 13 db may impose a limit to the minimum required f.m. power. This threshold limit is indicated by the horizontal line on Fig. 9 marked 13 db threshold.

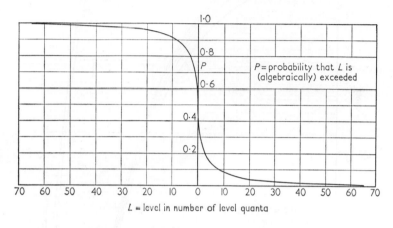

FIG. 10. Distribution of typical speech voltage.

To determine the required satellite r.f. peak power from Figs. 8 and 9: enter the range from ground to satellite on the axis of abscissae and the information bandwidth on the axis of ordinates. The required r.f. peak power in the satellite is then read from the oblique grids of constant power lines. In an f.m. system transmitting a stack of f.d.m. channels, as proposed for Fig. 9, to a close order of approximation, the received noise on each channel in the stack is proportional to the square of its mid-band frequency. Also the noise on the channel of highest mid-band frequency is substantially 3 times the mean noise per channel averaged over the whole stack of channels: i.e. 5 db worse. The speech to noise ratio in Fig. 9 is the mean noise per channel as defined in the last sentence. The consequence is that the speech to noise ratio on the worst channel is only 32 db. P-emphasis may, however, be introduced to bring the speech to noise ratio on all channels to substantially 37 db.

d. P.C.M. Transmission (Figs. 12 and 13) Figures 12 and 13 show the peak power required in the satellite for the system described in the body of the paper with the modification that an extra time unit is added in each time quantum to allow for a step in the process of p.c.m. coding and decoding made necessary by the nature of the code used.

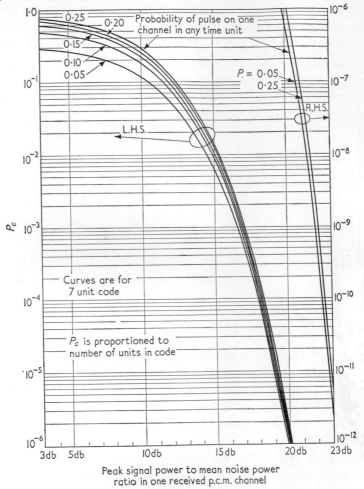

FIG. 11. P_c = probability of error in receiving a code character.

For level quantizing purposes the speech voltage is supposed to be limited so that all voltages which have a probability of less than 1% of being exceeded are suppressed. The residual voltage range is then quantized into a series of equal or tapered level steps.

The tolerable contribution to mean character error rate by receiving end noise is assumed to be 1 in 10^6, which fixes the required peak pulse to mean noise power ratio at 20 db, see Fig. 11. This gives a level error on the average once in 125 sec (i.e. once in $10^6/8000$ sec since there are 8000 time units and so 8000 characters per sec). Figure 11 is a plot of $p_c = $ the probability of failure of a character in the case of a 7 unit code against the ratio of peak pulse power to mean noise power for different values of $p = $ the probability of a pulse occurring in any time unit on any one telephone channel. Evidently p depends on the distribution of speech levels, the form of level quantizing and the type of code used.

The assumed permissible mean rate of overload of the satellite transmitter is once in 10^6 time units, and since there are 6 or 7 time units per time quantum and 8000 time quanta per sec, the mean rate of overload is either once in 20·8 sec or once in 17·8 sec.

The powers given in Figs. 12 and 13 are calculated on the assumption that the number of pulses in any one time unit which the satellite is designed to handle is the number which has a probability of 1 in 10^6 of being exceeded.

The voltage utilization of the satellite transmitter at any instant is assumed to be equal to the sum of the pulse voltages in the individual p.c.m. telephone channel information bands. This is pessimistic: see paragraph 5 of Conclusion.

Synchronous detection of the received basic p.c.m. information bands is assumed.

e. P.C.M. Transmission (*Fig. 12*) The p.c.m. code in this case differs from that used in Table II and Fig. 13 in that instead of using a minimum satellite-loading code, a code is used in which the whole level range of the speech is divided into individual level quanta, such that the probability of a level appearing in each level range is the same.

With this form of level quantizing the satellite loading is independent of the form of code used, and it is possible to choose a code which is most simple to implement in terms of electronic equipment. The probability p of a pulse appearing in any time unit is 0·125.

The basic bandwidth (= 1 sideband plus vestigial sideband) is assumed to be equal to the Nyquist band (half the number of bits per sec) multiplied by 2·5, see page 36.

f. P.C.M. Transmission (*Fig. 13*) The p.c.m. code in this case

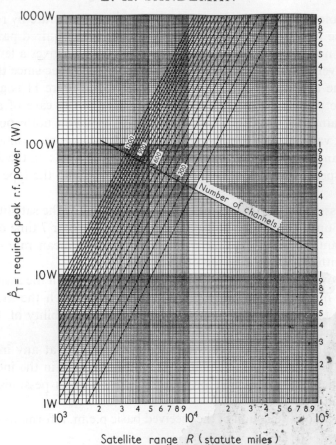

FIG. 12. P_T is the required peak r.f. power of single transponder in satellite in watts. With pulse code modulation 6 bit code equal probability level quanta.

1. P.C.M. band = 28 kc/s: 2. Received peak pulse to mean noise power ratio in p.c.m. channel = 20 db = probability of character error = 1 in 10^6: 3. Probability of transmitter overload in any time unit = 1 in 10^6: 4. Probability of a pulse on one channel in any time unit = 0·125: 5. Noise temperature 39°K: 6. Ground receiver dish = 100 ft diameter: 7. Satellite aerial gain 0·5.

approximates closely to the 7 unit power economy code shown in Table II. The basic bandwidth is as for Fig. 12.

To simplify computation, the code in Table II was modified by re-dividing each level range constituted by a number of level quanta using the same number of pulses per code (e.g. the level range from + 1½ to + 4½ or the level range from − 1½ to − 3½: all the levels in these ranges use 1 pulse per code) so that the new level quanta have an equal probability of occurrence.

This means that each group of codes having the same number of pulses refers to the same group of level ranges as in Table II, but within each group a redivision of the total level range into level quanta has been made.

This gave rise to a code with a probability of occurrence of a pulse in each time unit equal to 0·04. As a representation of the performance of the power economy code, Fig. 13 is evidently slightly pessimistic.

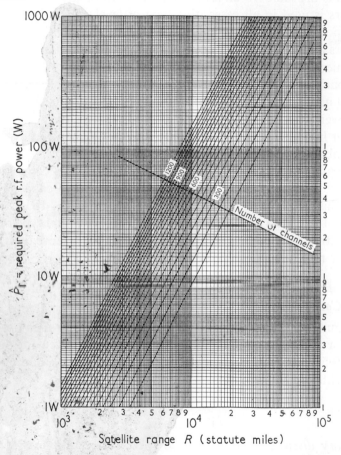

FIG. 13. P_T is the required peak r.f. power of single transponder in satellite in watts. With pulse code modulation 7 bit minimum power code.

1. P.C.M. band = 28 kc/s: 2. Received peak pulse to mean noise power ratio in p.c.m. channel = 20 db = probability of character error = 1 in 10^6: 3. Probability of transmitter overload in any time unit = 1 in 10^6: 4. Probability of a pulse on one channel in any time unit 0·04: 5. Noise temperature 39°K: 6. Ground receiver dish = 100 ft diameter: 7. Satellite aerial gain 0·5.

2. *Improvement due to Use of Power Economy Code*

The calculations of peak power in Fig. 13 are made for the code of Table II for the case where an extra time unit (making 8 in all per time quantum) is added for the purpose of coding and decoding, the level ranges being modified as just described.

As already indicated Fig. 12 gives the power using a level step grading such that the probability of a speech level occurring in any level quantum is the same for all level quanta. Provided the same set of code characters is used in all cases, this type of level quantizing makes the satellite power consumption independent of the distribution of the code characters among the level quanta. Such a condition of affairs provides a convenient reference of performance.

Comparison of the peak power demands shown in Figs. 12 and 13 for equal numbers of channels and equal satellite heights shows that the 2 bit power economy code used in Fig. 13 requires only a *fifth* of the r.f. power handling capacity in the satellite that is required by the 6 bit equal-probability code of Fig. 12.

For equal numbers of bits, therefore, the power economy code requires substantially a sixth of the power handling capacity in the satellite that is required by an equal probability code. It is shown in Appendix II that the power required by the conventional type of p.c.m. code resulting from coding with a digital voltmeter is more than 5 times the power required by the power economy code.

3. *Typical Values of Required Satellite R.F. Power using Economy P.C.M. Code*

The figures in Table VI are taken from Fig. 13, with small interpolation errors, for various heights of satellite. Figures are given for an aerial gain of 0·5 and for the gains shown on Fig. 7 corresponding to the selected satellite heights.

C. Television

Preliminary discussion The proposed speech transmission system described above using the p.c.m. power economy code requires r.f. peak powers in the satellite as shown in Table VI. With these powers for the conditions specified, the peak pulse power to mean noise power ratio on 1 pulse in any one received p.c.m. channel before decoding is 20 db corresponding to a probability of a missed pulse of 1 in 7×10^6 and the probability of a missed character of 1 in 10^6. Each p.c.m. channel occupies a bandwidth of 70 kc/s.

TABLE VI. Required r.f. peak power handling capacity of satellite transmitter
(watts)
p.c.m. + *f.d.m.* + *s.s.b.*
7 bit power economy code

Height of satellite		3000	6000	8000	22,300	statute miles
Range from satellite to edge of coverage area		5076·8	8477	10,619·6	25,279	statute miles
For gain = 0·5 power ratio	600 channels	12·1	33·7	53·1	300 W	
	900 channels	22·6	61·8	99·0	560 W	
	1200 channels	36·3	101·0	159·2	900 W	
For gain as in Fig. 7	600 channels	2·8	3·87	4·2	5·2 W	
	900 channels	5·3	7·1	7·9	9·7 W	
	1200 channels	8·53	11·6	12·6	15·6 W	
Gain in db from Fig. 7 for $\psi = 10°$		3·2	6·4	8·0	14·6 db	
Figure 7 gain as power ratio		2·13	4·4	6·3	28·8 power ratio	

For the p.c.m. economy code the probability of a pulse occurring in any one telephone channel is 0·04, and the number of pulses passing through the satellite in any one time unit due to all channels evidently depends on the number of channels. Particular examples are given in Table VII below.

TABLE VII

No. of telephone channels	Net frequency band-width occupied (Mc/s)	No. of pulses with a 1 in 10^6 probability of being exceeded
600	42	46·8
1000	70	69·4

The net frequency band occupied neglects frequency spacing between supergroups of telephone channels.

1. Method of Transmitting Television and Accompanying Speech Channel

It is proposed that the television band and the speech band should be p.c.m. coded and stacked side by side in the frequency domain. In the case of the speech band no frequency band compression will be used, and since the p.c.m. coding of speech and the stacking in the frequency domain has been already dealt with, this will not be considered further.

For TV the case without band compression is discussed in some detail below.

TABLE VIII

	System (No. of lines)	405	525	625	I.B.T.O. 625
f''_3	= Highest essential video frequency (Mc/s)	3	4·2	5	6
f''_2	= Lowest zero amplitude video frequency (Mc/s)	3·5	4·5	5·5	6·5
S	= Sampling rate (millions per sec)	7·5	10·5	12·5	15
$6·5\,S$	= TV frequency band (Mc/s) (easy filters)	48·75	68·25	81·25	97·5
$5\,S$	= TV frequency band (Mc/s) (special filters)	37·5	52·5	62·5	75

2. TV Transmission without Band Compression

For this purpose it is proposed to use p.c.m. The TV video wave is sampled at a rate S equal to twice the lowest frequency at which all amplitudes are substantially zero, or at $2\frac{1}{2}$ times the highest useful information frequency, whichever is the greater. With a 5 unit code and equal level steps some degradation of transmission can be observed due to quantization which disappears on going to a 6 unit code with equal level steps. It is more than likely that by using an optimized taper on the level steps it will be possible to achieve acceptable transmission with a 5 unit code, and a 5 unit code is assumed here.

The essential p.c.m. information band (after p.c.m. coding) is therefore $\frac{1}{2} \times 5 \times S$, and assuming vestigial sideband transmission, the frequency band occupied by the radiated TV signal is $2\frac{1}{2}$ times this Nyquist band, on the assumption of reasonably easily realizable filters. This gives a band occupancy of $6·25\ S$. With rather more difficult design of filters the band occupancy might be reduced to $5\ S$.

Table VIII gives relevant data for 4 standard TV systems. These are derived from Ref. 17.

a. Bandwidth Requirements Comparing the bandwidths with the bandwidths required for speech channels, and assuming, as seems desirable, that special filters should be used to conserve bandwidth, the 405 line system can be transmitted with the bandwidth required for 600 channels, while the I.B.T.O. 625 line system can be transmitted with the gross bandwidth required for 1000 channels, the gross bandwidth including frequency spacings between supergroups.

b. Satellite R.F. Power Requirements Provided the TV bandwidth requirement does not exceed that for the speech system, when the power in the satellite is adequate for the speech system as defined and designed above, it is more than adequate for the TV system. This may be checked by calculating the peak pulse power to mean noise power ratio on the received p.c.m. pulses in the TV case.

c. Example of Calculation of TV Peak Pulse Power to Mean Noise Power Ratio (*on received p.c.m. pulses*) The powers in Table VI have been calculated on the assumption that the number of pulses contributed by all channels never exceeds the numbers of pulses in Table VII and also on the assumption that the total voltage utilization of the satellite transmitter is proportional to the number of pulses, so that the power utilization is proportional to the square of the number of pulses. As a consequence the peak power in a received television pulse (allowing for 1 speech channel) can be at least as great as that corresponding to the numbers of pulses in Table VII less 1: in other words $45 \cdot 8^2 = 2098$ and $68 \cdot 4^2 = 4679$ times the power in 1 pulse in any one p.c.m. speech channel.

The received noise power accompanying any received TV channel pulse before p.c.m. decoding is evidently $5S/0 \cdot 07 = 71 \cdot 4 \, S$ times the noise power on any 70 kc/s $= 0 \cdot 07$ Mc/s wide p.c.m. speech channel at the receiving end before p.c.m. decoding, where $5 \, S$ is taken from Table VII, the special filter case being preferred, see Table VII.

If the 600 channel system is used to relay the 405 line system, the peak pulse power to mean noise power ratio for the TV system is larger than that on the speech system (which is 20 db) in the ratio $2098/71 \cdot 4$ $S = 2098/71 \cdot 4 \times 7 \cdot 5 = 3 \cdot 9$ times, or nearly 6 db.

If the 1000 channel system is used to relay the I.B.T.O. 625 line system, the peak pulse power to mean noise power ratio for the TV system is larger than that in the speech system in the ratio $4679/71 \cdot 4 \, S$ $= 4679/71 \cdot 4 \times 15 = 4 \cdot 3$ times, or just over 6 db.

In each case therefore the peak pulse power to mean noise power ratio is substantially 26 db. Referring to Fig. 11, for such a ratio, the probability of a character error on a 7 unit code is less than 1 in 10^{12}. On a 5 unit code the probability is even less. This means that with a sampling rate of even 15 million per sec, the mean error rate on picture elements is less than 15 per million sec.

d. Effect of Power Economy Code in TV Case The above calculations are valid for one TV channel whether a p.c.m. economy code is used for TV or not: the introduction of an economy code merely reduces the number of TV code pulses and does not change their amplitude. It therefore reduces the mean power loading of the satellite transmitter, but does not change the peak r.f. power.

e. Method of P.C.M. Coding In Ref. 24 Mr R. L. Carbrey of the Bell Telephone Laboratories describes an experimental TV p.c.m. system. A TV signal of nominal bandwidth 4·5 Mc/s is sampled 10 million times per sec and coded by means of a cathode ray tube beam deflected by the TV signal voltage. The beam is wide enough to embrace 6 positions on a target or screen in which a 6 bit code, corresponding to the level deflection, is written by means of holes in the screen. Six collector wires, 1 behind each hole, pick off the 6 digits and transmit them in parallel over 6 channels. There appears to be no difficulty at all in converting these parallel digits to a series group with a bit rate of 6×10^7.

3. TV Transmission with Band Compression

At least one system has been partially built and demonstrated (see Ref. 13) which gives a worthwhile amount of band reduction while providing a very presentable picture. The broad principle on which this system operates is to divide the video information band into 2 parts: 1 extending from zero frequency up to some frequency which has been instanced as 350 kc/s, and which provides information on the illumination of comparatively large areas: and 1 extending from 350 kc/s up to the highest video essential frequency (see Table VIII), which provides information on the edges or brightness transitions.

It also claims that a pulse code defining 128 levels requires about 12 million bits per sec to provide pictures of a 4 Mc/s TV system of the quality shown. Assuming vestigial sideband transmission, this requires a bandwidth of about 15 Mc/s.

A very rough statement of the method of transmitting the edges is that each frame (2 fields) is stored and analysed to determine the positions and magnitudes of the edges. Code groups are formed for each run of constant brightness which describe the length of the run, its brightness and the magnitude and sign of the transition at the end of each run. The resulting information is transmitted at a constant bit

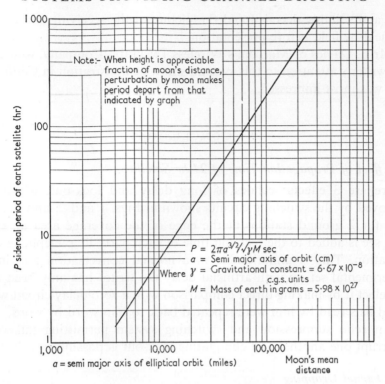

FIG. 14. Period of earth satellites.

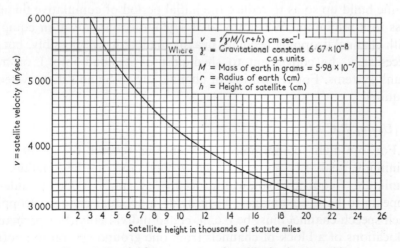

FIG. 15. Inertial frame velocity of satellite in circular orbit.

rate and suitably decoded and the picture at the receiving end is built up field by field before presentation.

Mr E. Davies of Marconi's Wireless Telegraph Company, who drew the writer's attention to this system, saw a demonstration in California and was most impressed.

VI CONCLUSIONS

1. Effect of Propagation Time of 24-hr Satellite

There are 3 effects: echoes of long delay and lock-out, which are removed by the switching circuits shown in Figs. 1 and 2, and the fact that for 2 ground stations at the edge of the coverage area an extra 0·56 sec is added to the reaction time of each party in a telephone conversation. This is unnoticeable in a normal conversation and in an experimental set up with artificial delay the writer has only been able to detect it by making direct comparison with a zero delay circuit when word lists read by him were repeated back to him word by word.

An echo suppression and switching system permitting talkers to interrupt one another and eliminating lock-out is possible.

2. Channel Dropping

It is considered that channel dropping is an essential feature of early satellite communication systems, in order to realize the maximum rate of traffic build up so as to reduce the initial period of cumulative financial loss to a minimum. When any area satellite communication equipped with channel dropping becomes heavily loaded, it will possibly, but not necessarily, be convenient to shed part of its load into 1 or more trunk systems. For realization of channel dropping an f.d.m. system is required.

3. Use of P.C.M.

For optimum compromise between the several requirements of minimum peak r.f. power on the ground and in the satellite, and minimum frequency bandwidth occupancy, pulse code modulation appears to have advantages. To provide individual channel dropping, each speech channel must be p.c.m. coded individually. For permanent allocations of a block of channels from one ground station to another, p.c.m. coding of a block of channels may be used. Reduced frequency bandwidth occupancy and some saving on r.f. peak power may be realized by individual p.c.m. coding of each speech channel of the block and the production of short pulses which are stacked side by side in time to constitute a sub t.d.m. system. The last two sentences apply equally to a trunk system.

4. Power Economy Code

A method of special p.c.m. coding has been indicated which results in minimum r.f. peak power requirements both on the ground and in the satellite.

Since writing the above the writer has found Ref. 21 which suggests the possibility of a power economy code.

5. R.F. Peak Power in the Satellite

The required r.f. peak power in the satellite is given in: Fig. 8 for s.s.b. all the way, Fig. 9 for f.m. all the way, Figs. 12 and 13 for p.c.m. all the way. Table VI summarizes p.c.m. satellite power requirements for particular cases.

In each case the same type of transmission is used from ground to satellite and from satellite to ground. In the s.s.b. case an f.d.m. stack of speech frequency bands is transmitted on s.s.b. In the f.m. case an f.d.m. stack of speech frequency bands is transmitted on f.m. In the p.c.m. case an f.d.m. stack of p.c.m. bands (individual or block bands) is transmitted on s.s.b. The f.m. case does not permit channel dropping.

The basis of peak r.f. power computation for the p.c.m. case is pessimistic since it has been assumed that the peak voltage due to n pulses (e.g. 48 or 69 pulses, see Table VII), is n times the voltage due to 1 pulse. While this is true if we wait long enough for the peak, on a statistical basis it is permissible to assume some value of peak voltage which has a permissible probability of being exceeded. This point requires investigation and it is expected that it will be found possible to reduce the satellite peak power requirement by a substantial factor. See Refs. 20, 22 and 23.

6. R.F. Mean Power in Satellite

F.M. Power

The mean r.f. power on f.m. is equal to the peak r.f. power given in Fig. 9.

P.C.M. Power for 7 Bit Minimum
Power Code of Fig. 13

From Table VII, the number of pulses which the satellite transmitter has to be designed to handle in the case of 600 channels, is 47. The maximum required power handling capacity is therefore 47^2 times the power in 1 pulse. From Fig. 13, the mean probability of a pulse occurring on any channel is 0·04. The mean power is therefore $0·04 \times 600 = 24$ times the power of 1 pulse.

The mean r.f. power is therefore equal to the peak r.f. power (as defined by the radio engineer) divided by $47^2/24 = 92$. Similarly for 1000 channels, the peak r.f. power must be divided by $69^2/40 = 119$.

The power read from Fig. 13 must therefore be divided by 92 for 600 channels and by 119 for 1000 channels to give the mean power.

Example: 600 channels 4 kc/s wide occupying 2·5 Mc/s on f.d.m.
    Satellite range 8000 miles.
    Satellite aerial gain = 0·5.

Power Requirements and Bandwidth

	Required r.f. power handling capacity (W)	Mean r.f. power (W)	Bandwidth occupied (Mc/s)
F.M. case	25 (Fig. 9)	25	105
P.C.M. case	30 (Fig. 13)	0·326	44·45

If the swing of the f.m. system is reduced to make it occupy the same bandwidth as the p.c.m. system, the mean power required for the f.m. system is 478 times that required for the p.c.m. system.

The mean r.f. power required in a 24-hr satellite, for the proposed p.c.m. system providing 600 channels, with an aerial gain of 28·8 (power ratio), is very nearly 60 mW.

7. Bandwidth Occupancy

A basic p.c.m. band occupies 70 kc/s; a basic group of 12 channels occupies 840 kc/s; a basic supergroup of 60 channels occupies 4200 kc/s.

The spacing between the 2 lowest supergroups is 12 Mc/s and between the remainder of the supergroups is 8 kc/s.

The bandwidth occupied by N channels is therefore $70N$ kc/s plus the spacings between supergroups.

Example

600 one-way channels occupies a bandwidth of $70 \times 600 + 2 \times 210 + 7 \times 140$ kc/s $= 43\cdot4$ Mc/s. Four such bands are required for 600 two-way channels: 2 contiguous bands from ground to satellite ($86\cdot8$ Mc/s); 2 continuous bands from satellite to ground ($86\cdot8$ Mc/s); with 2 kMc/s transmission frequency a spacing between the above 2 pairs of bands is required equal to 100 Mc/s. The total frequency band occupancy is $273\cdot6$ Mc/s.

A spacing of 100 Mc/s between bands has been assumed since it is considered that radio frequency of 2 kMc/s will be feasible, in which case such spacing will be adequate. For higher radio frequencies a larger spacing is required: this therefore constitutes an additional argument for the use of 2 kMc/s.

The bandwidth occupied by the f.m. system with a swing of ± 50 Mc/s is 100 Mc/s per leg or over 500 Mc/s for 4 legs. While the bandwidth occupied by the s.s.b. system is very much less, it makes unrealizable demands on power in the satellite.

8. The Penalty of Channel Dropping

The requirement of channel dropping restricts p.c.m. coding of complete standard C.C.I.T. or G.P.O. f.d.m. stacked speech bands. Where 2 ground stations require to use enough channels to occupy a particular group, supergroup or several supergroups, then the corresponding standard C.C.I.T. f.d.m. bands can be p.c.m. coded *en bloc*. When the number of channels permanently allocated between ground stations is smaller than a group (12 channels) it may be more convenient to provide for individual p.c.m. coding and decoding of channels. If ever the demand for individual switching of telephone channels from one ground station to another occurs, then individual p.c.m. coding and decoding is essential.

It will be seen that the requirement for channel dropping only increases the complexity of the system appreciably when there are a large number of individually coded channels. In such case the penalty has to be accepted because no other method is known of achieving channel dropping with commensurate power on ground or in satellite, with equal satellite simplicity and with comparative bandwidth occupancy.

ACKNOWLEDGMENTS

The writer has received considerable help and information from a number of people both in industry and in public services on both sides of the Atlantic and he would like to express his great appreciation of the very considerable kindness he has received.

Mr B. G. Anderson who carried out the extensive and detailed calculations which obtained the results disclosed in Part III cannot be let off so easily, and it is only fair to the writer to say that Mr Anderson refused to share the authorship of this paper. Without Mr Anderson's work and also that of Mr J. E. Fox, who checked the results, this paper would evidently have been impossible. Finally, the writer owes a very great deal to the speedy and extensive information service provided by the English Electric Aviation library staff under Mr T. M. Aitchison, whose digests on communications satellite activities have proved invaluable.

REFERENCES

1. HOLBROOK, B. D., AND DIXON, J. T. Load rating theory for multi-channel amplifiers. *Bell System Tech. J.*, p. 624, Oct. 1939.
2. RYDE, J. W. Attenuation and radar echoes produced at centimetre wavelengths by various meteorological phenomena. Report of a Conference on 8 April 1946, of the Physical Society and the Royal Meteorological Society.
3. LAW, H. B., REYNOLDS, J., AND SIMPSON, W. G. Channel equipment design for economy of bandwidth. *P.O. Elect. Engrs. J.*, p. 112, July 1960.
4. BROTHERTON, M. Amplifying with atoms. *Bell Labs. Record*, p. 163, May 1960.
5. PIERCE, J. R. Satellite systems for commercial communications. *IAS Paper No. 60–40* presented at the IAS 28th Meeting, New York, 25 Jan. 1960.
6. *Bell Labs. Record*, p. 433, Nov. 1960.
7. CHAFFEE, J. G. The application of negative feedback to frequency modulated systems. *Bell System Tech. J.*, p. 404, 1939.
8. BRAY, W. J. The standardization of international microwave radio-relay systems. *Proc. Instn Elect. Engrs.*, p. 180, Mar. 1961.
9. C.C.I.R. Documents of the IXth Plenary Assembly, Los Angeles. Vol. I, Dec. 1959.
10. C.C.I.F. green Book on Line Transmission Maintenance, XVIIIth Plenary Assembly, Geneva, Dec. 1956.
11. I.T.U. Radio Regulations, Geneva, 1959.
12. PRATT, H. J. Propagation, noise and general systems considerations in earth–space communications, *Trans. Inst. Radio Engrs.*, **CS-8**, No. 4, Dec. 1960.
13. SCHREIBER, W. F., AND KNAPP, C. F. TV bandwidth reduction by digital coding, Inst. Radio Engrs. National Convention Record part 4, p. 88, 1958.
14. COOMBES, W. C. National Bureau of Standards Technical Note No. 25, Communications theory aspects of television bandwith conservation.

15. GRAHAM, R. E. Subjective experiments in visual communication, Inst. Radio Engrs. National Convention Record part 4, p. 100, 1958.
16. BRADLEY, J. A PCM coding and decoding system. Symposium on Communications Satellites, 12 May 1961.
17. C.C.I.R. Documents of the IXth Plenary Assembly, Los Angeles, Vol. III, 1959.
18. *Aviation Week*, p. 79, 6 February 1961.
19. SLACK, M. Probability distribution of sinusoidal waves combined in random phase, *J. Inst. Elec. Engrs.*, **93**, part 3, 76, 1946.
20. CARBREY, R. L. Video transmission over telephone cable pairs by a pulse code modulation, *Proc. Am. Inst. Elec. Engrs.*, 1546, Sept. 1960.
21. FANO, R. M. The transmission of information, Technical Report No. 65 of Massachusetts Institute of Technology Research Laboratory of Electronics.
22. HAMILTON, B. P. Peak voltages in carrier telegraphy, *Bell Labs. Record*, Vol. XIX, No. 12, August 1941.
23. ANDERSON, D. R. Minimization of Maximum amplitude in Frequency multiplexing, *Proc. Inst. Radio Engrs.*, **49**, No. 1, 357, Jan. 1961.
24. CARBREY, R. L. Video transmission over telephone cable pairs by pulse code modulation, *Proc. Inst. Radio Engrs.*, **48**, No. 9.
25. An inquiry into the allocations of frequency bands for space communications, Depositions of American Telephone and Telegraph Company before the Federal Communications Commission, Washington 25 D.C.
26. Depositions of the Radio Corporation of America, as Ref. 25.
27. E.M.I. cathode ray tube decoding devices for digital communications systems, S. C. Ghose. URSI Symposium on Space Communications, Paris, 1961.

APPENDIX I

Definitions

Channel or Communication Channel

A means of conveying one message at a time whether in the form of speech, telegraphy, telex, television or facsimile transmission. A channel may be one-way, in which case it can only transmit information from a point A, for instance to a point B. A two-way channel can on the other hand transmit also from point B to point A. It is usually left to the context to decide whether a channel is one-way or two-way.

2-Wire Circuit

A channel, which may be one-way or two-way, constituted by a single pair of wires.

4-wire Circuit

A two-way channel constituted by 2 pairs of wires, 1 pair for transmission in one direction and 1 pair for transmission in the other. In telephony each pair of wires of a 4-wire circuit normally carries the half of a telephone conversation going one way.

Trunk or Trunk System

A communication system in which all channels start at one terminal and end at one other terminal.

Area Communication System

A system in which any ground station in a specified communication area can communicate with any other ground station either in that area, or in a suitable defined area.

Multiplex

Any means of superimposing a number of channels on a single means of communication such as a single radio transmitter or receiver, a co-axial cable, a single circuit etc.

F.D.M. = Frequency Division Multiplex

A system of multiplex in which a number of information frequency bands are stacked side by side in the frequency domain, usually as close as the filter attenuation slopes will permit.

T.D.M. = Time Division Multiplex

A system of multiplex in which a single communication means is shared among a number of communication channels by stacking elements of information from each channel sequentially in time. In this paper pulses which for single channel working would have a duration of 1 time unit (see below) are divided in duration by the number of channels which it is required to superimpose, and during each time unit shortened pulses are transmitted in turn from each channel, the same process being repeated during each time unit.

A.M. or F.M. = Amplitude or Frequency Modulation respectively

The systems of modulation in which respectively the amplitude or frequency of a carrier wave is made proportional to the instantaneous amplitude of a voltage carrying information whether from 1 channel or a number.

S.S.B. = Single Sideband (Transmission)

The system of transmission in which an a.m. wave is processed to extract 1 sideband which is then transmitted.

Vestigial Sideband Transmission

When a modulating information band of frequencies extends down to zero frequency (such as the band of frequencies resulting from p.c.m. coding a speech wave), even using double balanced modulators it is not possible to realize s.s.b. transmission by extracting 1 sideband, completely free of the other. In such cases 1 sideband and a "vestige" of the other is selected, suitable equalization being provided to give an overall flat frequency response after detection. In the f.d.m. system discussed here the resulting band of frequencies constituted by the selected sideband and vestigial sideband is stacked in an array of similar frequency bands as close as is permitted by the relevant system of multiple modulation used for the purpose.

P.C.M. = Pulse Code Modulation

In this system time and the information voltage level are independently quantized. During each time quantum the voltage level is sampled and during the next time quantum a coded train of pulses is transmitted to define the voltage level. The train is decoded at the receiving end.

Group

In an f.d.m. system this refers to a group of channels selected for mutual association in a *basic group* so located in the frequency domain that stacking of groups in the frequency domain by transfer of the basic group from one frequency band to another is facilitated. The G.P.O. use groups of 12 and 16 channels where frequency stacking is used without p.c.m. coding. With no particular emphasis it is here proposed to use groups of 12 channels as a basis for discussion.

Supergroup

This contains 5 groups (q.v.) stacked initially in a basic supergroup located in a frequency band convenient for stacking of supergroups.

Assembly

An array of information bands stacked in the frequency domain regardless of how derived.

Sampling Frequency

The number of times per sec that the voltage level of an information band of frequencies is sampled,

Time Quantum

The reciprocal of the sampling frequency equals the duration of time occupied (*a*) by a sampling process (*b*) by the transmission of a code character consisting of a series of time units either occupied or not occupied by pulses.

Time Unit

The element of time during which one element of a character defining a level is transmitted. The duration of such a time unit is $1/n$ of the duration of a time quantum where n is the number of time units in one character.

Character

The combination of time units occupied by pulses and blank time units used to define or describe a voltage level.

Appendix II

Comparative Satellite R.F. Powers Required with Different Codes

Comparison of Equal Probability Code with Conventional Binary Digital Code Normally Used for P.C.M.

The term binary digital code is used here to indicate the code which results from the use of a digital voltmeter to measure the voltage level and originate a binary code to indicate the level. It is assumed that the voltmeter represents the most extreme negative level by zero, the next level by binary digit 1 and the next level by binary digit 10, the next by 11 and so on. The most positive level will then be represented by a binary number with a 1 in every place, e.g. for a 7 bit code, the number representing the highest positive level is 1111111.

Under such conditions, for a 7 bit code defining 128 levels, if the first 128 binary numbers are written down in order it will be found that in the first digit, there are alternate 1's and zeros, while in the highest digit (representing 2^6) there are no 1's in any of the numbers representing all the negative levels, but that all the numbers representing positive levels have 1's in the 2^6 digit place.

It will be immediately evident that the probability of occurrence of a digit in the highest digit place is 0·5 if the speech is continuous. In practice, even with 1 subscriber talking continuously, the probability is only 0·25 due to pauses, while in normal conversation the probability is 0·125. This is the same probability that occurs in every digit place (and so in every time unit) with the equal probability code. It is also nearly true that the probability of occurrence of a digit in the lowest digit place is 0·125. This is because the digits alternate with zeros in the digits place, while the difference in probability of occurrence of each linear quantized level from one level to the next is not high.

The probability of occurrence of a pulse in any time unit other than that corresponding to the digits place or the highest power of 2 place, with a binary digit code as just defined, is less than 0·125. The probability of occurrence of a pulse in every time unit of an equal probability code as defined earlier in the text is 0·125.

The consequence is that the frequency of overload of a transmitter set to handle any number of pulses occurring simultaneously during 1 time unit, is less for the binary code with linear quantization than for the equal probability code.

It is possible to show, however, that the difference in required r.f. power handling capacity in the satellite for the 2 cases is negligibly small, although a full analysis of the situation has not been made on account of the complexity of the situation.

Figure 16 shows a plot of the probability p_N that any number N of pulses in any time unit is exceeded, when there are 600 channels each contributing pulses in the same time unit, and the probability of a pulse occurring in any channel in the time unit considered is 0·125.

Since the probability $p = 0·125$ applies to all time units of the equal probability code and to 2 time units of the binary digital code, if we confine our attention to the 2 time units in question, Fig. 16 can be used to provide information relevant to both codes.

Now assume that we have set up a transmitter so that the probability of overload in any one time unit with the equal probability code is 10^{-6}. Looking at Fig. 16 it is seen that the transmitter must be set up to handle 114 pulses since 113·5 pulses has no meaning: only integral numbers of pulses can be considered.

It will be noted that, with this setting of the transmitter, the contributions of each of the 7 time units to overloading with the equal probability code will each be once in 10^6 time units so that the total contribution of 7 time units is 7 times in 10^6 characters.

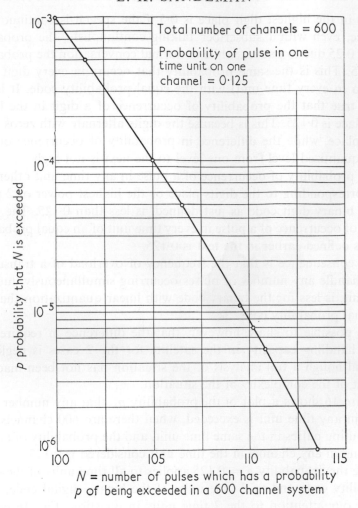

Total number of channels = 600

Probability of pulse in one time unit on one channel = 0·125

p probability that N is exceeded

N = number of pulses which has a probability p of being exceeded in a 600 channel system

FIG. 16

A useful question to ask now is to what pulse handling capacity can we reduce the satellite transmitter to make the rate of overload, contributed by the first and last time units only of each binary digit code character, equal to 7 times in 10^6 characters. The answer is evidently given by the number N of pulses corresponding to a rate of overload of 3·5 times in 10^6 time units, that is to a probability of overload of $3·5 \times 10^{-6}$.

(With the satellite transmitter designed to handle this number of pulses, the rate of overload with the binary digit code will in fact be slightly greater than 7 times in 10^6 characters owing to the contributions of the 5 time units which have been neglected.)

Entering the probability $3 \cdot 5 \times 10^{-6}$ in Fig. 16 gives $N = 111 \cdot 5$ pulses. Since half a pulse has no meaning, in order to favour the binary digital code, assume 111 pulses.

Since it is assumed that the pulse voltages add linearly, this means that the binary code requires a power handling capacity rather more than $(111/114)^2 = 0 \cdot 95$ times that of the equal probability code in order to give the same rate of overload. It will be evident that the figure of $0 \cdot 95$ is on the low side, not only because the other digit places of the binary digit code have been neglected, but also because the reading of the number of pulses required has been rounded off in each case to favour the binary digital code, and finally because Fig. 13 has been computed using an unfavourable modification of the power economy code to simplify computation.

Comparison of Power Economy Code with Binary Digital Code

The conclusion is that the power economy code which, for equal number of bits in the 2 codes requires substantially a sixth of the power handling capacity of the equal probability code, requires less than a fifth of the power handling capacity of the binary digital code as defined.

A P.C.M. CODING AND DECODING SYSTEM

J. BRADLEY

British Aircraft Corporation, Ltd., Luton, Bedfordshire, England

I. INTRODUCTION

In his paper on satellite communication (Ref. 1), Dr Sandeman suggests that, for speech transmission, in order to make the most efficient use of the available bandwidth and to minimize the transmitter power requirement, the speech voltage should be transmitted in terms of a set of specially coded binary numbers. The instantaneous voltage amplitude is measured substantially every 125 μ sec and described by a 6 or 7 digit word. The code is chosen so that as far as possible, the lower the speech level the more zeros are contained in the binary code representing it. In this way the average transmitted power is minimized.

This paper suggests a method by which the required coding and decoding may be achieved using any code. For the proposed type of code the method appears to offer a fairly simple, reliable and inexpensive solution, though it would not meet the speed requirement necessary for the transmission of video signals. Present methods for the p.c.m. of video signals adopt beam coding tubes (Ref. 2) and are restricted to the Gray code, though further development may overcome this limitation.

The method outlined below is tentative and has not been proved in hardware, though much of the circuitry is standard computer practice. It uses standard ferrite cores as memory units for coding and decoding, though there may be some advantage in using instead multi-apertured ferrite cores as described in Refs. 3 and 4.

II. POSSIBLE METHODS

The basic requirement is to determine the instantaneous signal amplitude in relation to a number of levels, here assumed to be 64, to decide upon the appropriate level and to open a gate which will pass the corresponding code word to the transmitter. Since the signal amplitude will be continually changing, the instantaneous level must be remembered during measurement. This can be done by storing the level on a capacitor.

The most obvious method is then to present the capacitor voltage to 64 comparators, the outputs from which would be used to open the appropriate gate from a word generator to the transmitter. The word generator would provide the 64 words and would be common to all channels. The gating signals would require to be stored on flip flops to ensure that they did not change during the read out of the selected word. Such a system would therefore require 64 comparators with associated logic for selecting the level, 64 flip flop stores and 64 gates.

An alternative method would be to use a single comparator and 6 flip flops controlling switched resistors as in a standard digital voltmeter. This would code the level in the normal binary code. Six further flip flops might be necessary to store the information during the new measurement unless the measurement took less than 18 μ sec. The outputs of the flip flops would pass into a diode matrix which would separate the 64 states and open the appropriate gate to let the required code through. The matrix would require about 384 diodes and this method may be more expensive than the earlier one.

A modification of this method would be to replace the diode matrix and the 6 flip flop stores by a matrix of square loop ferrite cores. The core memory could be used to store the output of the digital voltmeter and generate the required code from within itself. It is this method which is discussed in this paper.

III. OUTLINE OF THE METHOD

The block diagram is indicated in Fig. 1. The instantaneous input signal level is staticized on a capacitor during measurement by opening the isolating switch. The capacitor voltage is read by a 6-stage digital voltmeter of conventional form. The binary coded output is transferred to a magnetic core store which initially contains all the binary numbers from 0 to 63. When the voltmeter output is set into this matrix all but one number will be changed: the number corresponds to the voltmeter output. Sixty-four output wires from the store detect changes in the store content. One wire will detect no change and is therefore uniquely determined.

This wire is now used to excite a second magnetic store which will generate the corresponding code word. This code word is read out serially into the transmitter.

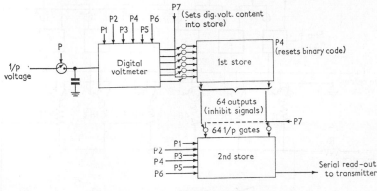

FIG. 1.

The first magnetic memory can be designed around 12 cores and the second requires 6 cores. Unfortunately, unless there is some helpful relationship between the standard binary code and the special binary code, the 64 interconnections between the two stores cannot be avoided. Each of these connections will require a drive amplifier and the economics of the method depend largely upon the simplicity of this unit.

Decoding the received signal follows a similar pattern.

The equipment required for the above purpose and its method of operation will now be described in detail.

IV. Sequential Pulse Generator

This generates 7 pulses P_1, P_2 ... P_7 in sequence during each sampling period of 125 μ sec. The generator provides pulses to all coding equipments at one ground terminal.

V. The Digital Voltmeter

This is a conventional unit, see Fig. 2. It consists of 6 flip flops each controlling a two-way switch. Resistors are connected to the poles of the switches and are switched to earth or to a negative reference voltage by the states of the flip flops. The resistor values increase by factors of 2 in accordance with the value allotted to the contents of the successive flip flops. The resistors are brought to a summing point together with the input resistor and a bias resistor from a positive reference

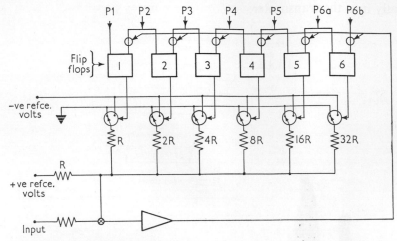

FIG. 2. Digital Voltmeter.

voltage. The voltage at the summing point is brought to zero using an amplified error signal to control the settings of the flip flops. The flip flops are set sequentially by signals from a pulse generator, beginning with the most significant flip flop. If the contribution from a resistor reverses the polarity of the error signal, the flip flop is reset on the next pulse. The final states of the flip flops indicate the amplitude of the voltage as a 6-digit binary number.

Such a device can operate in a much shorter time than required for this application. The requirement here is that the operating time plus the time during which the input capacitor is switched on the signal shall not exceed 300 μ sec.

VI. THE CORE STORES

Consider the array of 12 magnetic cores shown in Fig. 3. The cores are arranged in 2 rows of 6 cores. Each row is threaded by a "set" wire. A current pulse down the left-hand row will set all these cores into a state we may call "0". A current pulse down the right-hand row will set all these cores to a state we may call "1". The directions of the currents are indicated throughout by the arrows; the states are identified merely for the purpose of description.

The left-hand cores are linked individually via switches to the output of the sequential pulse generator providing pulse P_7. The right-hand cores are similarly linked to the same source. The 6 pairs of switches are controlled by the 6 voltmeter flip flops so that if a flip flop contains a "0" the left-hand switch is open and the right-hand switch is closed.

The cores are threaded with another 64 wires so that each wire passes through *one* core of each pair in a unique fashion. Each core is therefore threaded by 32 such wires. One wire is shown in Fig. 3. It represents the binary number 010011 and is therefore threaded through the first (i.e. top) left-hand (0) core, the second right-hand (1) core and so on.

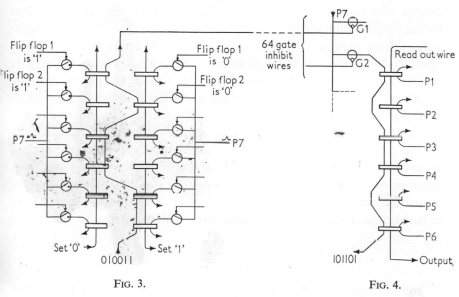

FIG. 3.   FIG. 4.

Now suppose the digital voltmeter flip flops register this same number. Then the first, third and fourth right-hand switches and the second, fifth and sixth left-hand switches will be closed. Therefore when the pulse P_7 is generated these cores will be changed to their opposite states. The resultant changes of flux will generate pulses in all wires except the one shown (i.e. 63 wires) since this will be the only wire not threading any of the switched cores. The output pulses from the 63 wires are used to inhibit 63 of the gates G_1 to G_{64}.

At the same time a second store of 6 cores is pulsed by P_7 through G_1, G_2 etc, and the output pulses from the first store inhibit the pulse P_7 passing into 63 of the 64 wires into the second store. This store is shown in Fig. 4.

The 6 cores are threaded by the 64 wires in an identical manner to the right-hand row of Fig. 3. Instead of each wire passing through one of a pair of cores, it either passes through the single core or not. One such wire is shown in the figure. This corresponds to the number 101101. The whole row of cores is sequentially reset to the "0" state by pulses P_1 to P_6 from the sequential pulse generator. During this resetting process a serial 6-digit word is induced in a single read out wire connected to the output.

Suppose all the cores are initially set to the "0" state. When the unique wire shown is pulsed by P_7 (the other 63 wires being inhibited by the outputs from the first store) the states of the first, third and fourth cores will be changed to "1". Therefore when the cores are sequentially reset to "0", pulses will be generated by these 3 cores in the wire to the transmitter. This resultant combination of pulses suitably amplified and shaped provides the coded 6-digit transmission signal. A sync pulse is inserted in the seventh digit position which destroys any spurious pulses generated by P_7. This completes the word cycle.

The required change from the standard binary code to the special binary code is achieved by suitable arrangement of the 64 interconnections between the 2 stores. Spurious pulses generated by resetting the first store, which may be done in any period other than P_7, cannot pass to the second store, since P_7 acts as a strobe signal.

VII. DECODING THE RECEPTION SIGNAL

The decoding process follows a similar pattern. The incoming signal word, excluding the sync pulse, is written serially, using the coincident current technique, into 6 pairs of cores so that if a "0" occurs in the signal the corresponding pair of cores is set to "0" and "1" respectively. Likewise if a "1" occurs the pair of cores is set to "1" and "0".

Figure 5 shows the wiring. The 12 cores are threaded by 64 wires as in Fig. 3 of which one is shown corresponding to a signal word 010011. The cores are pulsed in pairs as in Fig. 4 by pulses P_1 to P_6. The cores are threaded with 2 "set" wires which set both columns to a state called "0". In addition both rows of cores are threaded by a "write" wire from the receiver as shown. Signals down this wire will be positive current pulses if the signal digit is "1" and negative current pulses if the signal digit is zero. The arrows show the directions of

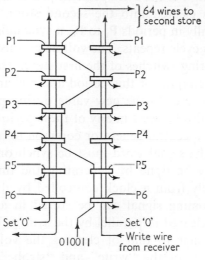

FIG. 5.

positive currents. The strobe currents P_1 to P_6 and the signal currents are both only half that required to change the state of a core. Thus both must occur together to be effective. Positive signal currents aid the strobe currents to set left-hand cores to "1" and negative signal currents aid the strobe currents to set right-hand cores to "1". In this way the signal word is set into the left-hand row and its complement is set into the right-hand row. The cores are then cleared by the "set" pulses, which are P_7 pulses, into a second store of 6 cores as in transmission. This store is thereby set to the binary equivalent of the signal word and is preferably cleared in parallel (i.e. in one pulse period, for instance by feeding pulse P_1 into all cores of the analogue of the 6-core matrix in Fig. 4), into 6 flip flops. The flip flops control two-way switches which set up 6 summing resistors to reproduce the voltage analogue of the binary word at the summing point output.

It is possible to avoid the requirement to generate positive and negative pulses from the incoming signal. Instead only the positive pulses may be generated and the right-hand strobe currents are then doubled. This method may run into difficulty with synchronization.

VIII. Timing

If signal amplitude is transmitted every 125 μ sec in a 7-digit word, the timing is as follows. Consider transmission first.

The seventh digit is the sync pulse in period P_7. During this period the first store is cleared into the second store. The second store is cleared sequentially in periods P_1 to P_6, a sync pulse being added to the signal in P_7. This cycle repeats. The voltmeter must be idle in P_7 since it controls the clearing switches of the first store. Therefore the capacitor is switched to the input for this period and recharged to the new level. The voltmeter is reset to the new value during periods P_1 to P_6. The first matrix can also be reset in any of these periods. The timing pulses are produced by a central generator common to all channels.

In reception the signal word must be synchronized with the timing pulses. This may be done by generating the timing pulses for each channel separately from a clock controlled by the sync signal. Alternatively the incoming signal can be stored in a transistor-core shift register. This is cleared in parallel by the sync pulse into the first store in a manner similar to that of clearing the voltmeter during transmission. The roles of the "write" and "strobe" wires in Fig. 5 are reversed and 6 driving amplifiers would be required on the shift register outputs.

Detailed examination of a complete multichannel system would be required to determine which method would be cheaper. Much depends on the ease of producing current pulses of many amperes and short duration, and how such generators could be shared among the various channels to even out the loading.

IX. Further Observations

The point made in the last paragraph applies to the whole system. At the beginning of this paper several alternative methods were outlined. All are fairly expensive when considered in detail. The virtue of the system described in the bulk of this paper depends much on how economically the common pulse generators and the 64 gated amplifiers required to drive between the 2 stores can be made.

The system also presents certain problems which can only be resolved by simple experiments. In particular there is the question of whether cross coupling between the 64 wires in the first transmission store would generate spurious signals large enough to be indistinguishable from a genuine pulse. If necessary this problem could be diminished by setting up the store sequentially since the voltmeter is inherently a sequential device. The drive amplifiers to the second store would then need to store any inhibit signals generated until the arrival of P_7. Such problems require practical investigation.

Finally, it may be possible to economize by using much of the transmission coding equipment for reception purposes. This however will complicate the already difficult problem of giving preference to one caller when both wish to speak.

REFERENCES

1. SANDEMAN, E. K. Satellite communication system providing channel dropping, British Interplanetary Society Symposium on Communication Satellites, 12 May 1961.
2. CARBREY, R. L. Video transmission over telephone cable pairs by pulse code modulation, *Proc. Inst. Radio Engrs.*, **48**, No. 9, September 1960.
3. RAJCHMAN, J. A., AND LO, A. W. The transfluxor, *Proc. Inst. Radio Engrs.*, **44**, 321–32.
4. RAJCHMAN, J. A., AND CRANE, H. D. Current steering in magnetic circuits, *Trans. Inst. Radio Engrs.*, March 1957.

SOME PRACTICAL PROBLEMS OF
SATELLITE COMMUNICATION

J. R. PIERCE

Bell Telephone Laboratories, Murray Hill, New Jersey, U.S.A.

I. INTRODUCTION

It has become clear that the problems of satellite communication are many.* It is certainly impossible to argue them to any conclusion in a half hour, or indeed, in many hours. The purpose of this article is to describe briefly some of the important practical problems so far encountered, and to state the author's views concerning them.

Before going on to this survey of individual problems, the author wishes to say that he believes that practical satellite communication can be achieved without any new ideas or "breakthroughs" (whatever they may be). In fact, several sorts of satellite communications systems could be made. To make any one sort work satisfactorily would take a great deal of time, money, research, and development. To choose sensibly among various general sorts of systems, one must think of such broad matters as: how quickly can a useful system be put into operation; how satisfactory will it be; is it an economical or a costly system? Such considerations lead to several narrower but still very broad practical problems. Some of these are mentioned later and commented upon briefly.

A first and vitally important problem is how satellite communication can be fitted into our already highly developed communications systems, so as to expand their usefulness. While a satellite communications system

* Satellite communication has been under serious consideration at the Bell Telephone Laboratories since 1954. A large and intensive programme aimed at practical satellite communication has been under way for over 2 years. The research department's part in Project Echo was the first substantial experimental undertaking. Work in the research department is continuing and increasing. The development departments of the Bell Laboratories have undertaken the responsibility for extensive experimental and development work leading to a useful active satellite communications system. Thus, we have been exposed to the problems of satellite communication through work over a period of years, involving many tens of people and several million dollars.

might conceivably serve some individual and self-sufficient purpose, such a narrow and restricted use would seem a poor basis for what will necessarily be a very expensive undertaking.

There are about 134 million telephones in the world. Particularly in Europe, North America and Japan, these are interconnected by complicated, versatile and adequate transmission and switching facilities. Together with these, the telephone operating companies or agencies provide facilities for the transmission of teletypewriter data, facsimile and television signals.

Transoceanic telephone traffic is growing rapidly. While there are a number of transoceanic telephone cables, and more will be built, the total number of transoceanic telephone circuits, cable and radio, is inadequate. We have no circuits of television bandwidth. It seems to me that the greatest, the most useful and the most economically attractive service that satellite communication could provide is that of linking together the large, highly developed communication systems of various continents.

What problems do we face in doing this? A pattern of international communication already exists. International operating standards must be compatible. Over a period of decades they have been worked out through the International Telecommunications Union, now a part of the United Nations. In each country the agency which provides telephone service owns and operates transatlantic radio terminals. International telephone cables are owned jointly by the agencies of the countries they interconnect. Thus, the cable connecting the United States and Great Britain is owned jointly and equally by the British Post Office and the American Telephone and Telegraph Company.

It seems to me that the quickest, the surest and perhaps the only way to achieve useful, mutually profitable satellite communication is within this established, functioning, non-political international framework of co-operative technical standards, ownership and operation. It seems to me that any other approach would create a host of new and unnecessary problems without providing any substantial benefits.

Another and narrower problem has to do with the particular technical standards which must be attained and maintained. There are established standards covering such matters as bandwidth and signal to noise, and much more detailed and recondite matters as well. Among these is a recommended practice that the overall one-way time delay of a telephone circuit shall be less than a quarter of a second. This is somewhat

PLATE 1. Satellite tracking station.

less than the propagation time to a 24-hr, 22,300 mile high stationary satellite and back to earth.

This allowable delay was set many years ago for what seemed wise reasons. Even using a 4-wire circuit, with entirely separate paths between each transmitter and receiver, the effect of this delay is noticeable in some conversations. In actual telephony, 2-wire circuits are used which cause an intolerable echo when the delay is so long. To overcome this, echo suppressors are used on long circuits; these interrupt transmission while a signal is being received. This can result in unfortunate deletions from or blocking of conversations.

The degradation inherent in the delay associated with stationary satellites must be carefully evaluated. It is clear, however, that a 24-hr satellite can never give us quite as good a telephone circuit as those we now have via cable.

The launching of satellites is a difficult and expensive matter. Experience has shown that at present new or modified launching vehicles are uncertain in their operation, and even vehicles which have been used many times are not unfailing. This, together with the high cost of launchings and the higher costs of developing a new vehicle, convinces me that the only economically practical way to launch a satellite is to use the best adapted, well tried, currently available vehicle.

Fortunately other space needs have given us vehicles suitable for the launching of communication satellites and further work will give us better vehicles, quite independently of any needs for satellite communication. We should take advantage of this fact rather than burdening satellite communication with extremely expensive and necessarily limited vehicle development programmes.

There has been much discussion of the tracking of satellites. It has frequently been argued that the fact that a stationary satellite need not be tracked will lead to a substantial economic advantage. It has even been proposed that stationary, non-steerable antennas might be used. I'd certainly hate to gamble on a 24-hr satellite staying in the beam of a stationary antenna in an experiment or even in an early and untried system.

Tracking is a difficult problem. Electronics is a sophisticated and powerful art. During the ECHO experiment, by using predicted positions, the ECHO satellite was tracked to $\pm 0.2°$, and sometimes to $\pm 0.1°$, when the data was fresh. This is not good enough for a commercial system using larger antennas, but better tracking can be attained. Further,

C.S.–D

when one considers actual costs, the provision for tracking will not be a major factor in the cost of ground terminals and it will be an even smaller factor in the cost of an overall system, including the cost of launching and replacing satellites. The cost of launching and replacing satellites will be the greatest cost in a satellite communications system.

This makes the life of satellites one of the most critical problems in a satellite system. A factor of a very few-fold in life could make or break a satellite system.

II. RADIATION PROBLEMS

Space poses many inadequately explored problems important to satellite life. Most active space payloads depend on solar cells, together with storage batteries, for electrical power. Sunlight has an energy of about 130 W per square foot, and solar cells can turn about 10% or 13 W of this into electric power initially.

Unfortunately, radiation pervades space. Cosmic radiation has little effect on solar cells. Less than a tenth of an inch of glass, quartz or sapphire will shield solar cells adequately from the electrons of the outer van Allen belt, which is most intense about 13,000 miles above the earth's surface. It appears impossible to protect solar cells against the effect of the protons of the inner van Allen belt, which is most intense at an altitude of around 2000 miles and which extends to an altitude of perhaps 5000 miles.

While the discovery and exploration of the van Allen belts is a major scientific feat, the extent, intensity, and composition of the radiation have not been carefully measured, nor their changes with time accurately determined. While we know how much degradation radiation of various energies causes in solar cells, our uncertainties concerning the nature and amount of radiation which a satellite will encounter are great. We are uncertain by perhaps a factor of 3 as to when the efficiency of a solar cell will have fallen to some preset design limit, say, 5%. Our best present estimate is that the solar cells in a satellite a few thousand miles high would last for from several months to several years. Solar cells on higher satellites would last longer.

While all solid-state electronic devices are subject to the hazards of radiation, transistors and diodes are in general less sensitive to radiation than are solar cells. Further, they are better protected, both by metal encapsulations and by their location within a metal-clad satellite structure. This does not mean, however, that satellite electronic equipment other than solar cells does not face special problems and hazards in space. All equipment must withstand tens of g's of acceleration during launch.

Ultimately, only radiation cooling is available. Heat must be conducted to radiating surfaces. There can be convection even in a pressurized container only if the satellite is spun, because there is no weight in an orbiting satellite. In an alternate environment of sunlight and shadow, the temperature variation of the exposed parts of a satellite will be large, though the thermal capacity of interior parts can reduce their temperature fluctuations.

Vacuum and gas at atmospheric pressure are good insulators but, during ascent, pressures are encountered at which a discharge can be initiated at a comparatively low voltage. Once in orbit, pressurized containers may leak. The seals of storage batteries sometimes leak. Vapours from apparatus can produce appreciable gas pressures in enclosed spaces even if these are not completely gas tight.

The dependability, the life of satellite electronic equipment, is of primary importance to satellite communication. The chief initial cost of a satellite communications system will be the cost of launching and replacing satellites. Unless the life of satellites is many years, the chief operating cost will be the cost of replacing satellites. The practicability of satellite communication depends on obtaining long life in the novel and hazardous environment of space.

Undoubtedly, long life can be attained in space. The VANGUARD transmitter is still functioning after 3 years in orbit. This is a very simple transistor oscillator, and enough solar cells were used to keep it going despite a drastic fall in their efficiency. More complicated satellite payloads have lasted as long as a year (an unsatisfactorily short life for a communications satellite), but many others have failed after a few weeks or a few months in orbit, and some have failed in whole or in part during launch.

The operation of apparatus on command from the ground has proved particularly hazardous. One would expect apparatus to function most reliably when it runs continuously without interruption, and experience indicates that this is indeed the case.

III. ORIENTATION PROBLEMS

So far, I have discussed only the problems of the minimum essential electronic equipment for an active satellite. This equipment could be enough for a communications satellite, because low satellites could be operated unoriented, or with very crude orientation. Of course, there are advantages in orienting any satellite, for this cuts down the power required in a low satellite appreciably, and that required in a high satellite greatly. However, orientation could cost more than it is worth if it resulted in decreased satellite life, and stationary satellites call for station keeping as well as for orientation.

What is the state of the art of orientation and station keeping? ATLAS nose cones have been successfully oriented for tens of minutes. DISCOVERER satellites have been oriented for days. If communications satellites are to be useful and to compete economically with cables, they must last for years.

Will gas valves operate for years without leaking? Will bearings operate for years in vacuum without freezing? What is the chance that during its years of operation some brief malfunction of orientation gear will set the satellite spinning? Just how dependable will orientation and station-keeping gear be? If such equipment depends on commands from earth, can we really count on its responding to our commands and on its not responding to signals of other origin? Today we have only conjecture. We will be able to answer such questions only on the basis of experiments and experience.

We might wish to orient a satellite in a low orbit. The magnetic field is very small at an altitude of 22,300 miles, and it changes with time, for it depends not only on the magnetism of the earth but on the flow of charged particles from the sun. At altitudes of a few thousand miles, the earth's field is strong enough to serve as a reference or to provide an appreciable force in orienting a satellite. But, this means of orientation has not been adequately explored experimentally.

We might make use of another force in orienting a low satellite. For a given angular velocity, centrifugal force increases with increasing radius of orbit, while the force due to gravity decreases with increasing radius. Imagine a satellite made up of 2 equal weights tied together with a wire and aligned radially with respect to the earth. The centrifugal force will be greater on the outer weight, and the force of gravity will be greater on the inner weight. There will be a small tension in the wire. If displaced, the system will tend to return to its radial orientation. But, this force is very small.

TIROS I and TIROS II, the weather satellites, have been spin oriented.

This keeps the axis of spin pointed in one direction. As a satellite rotates in the earth's magnetic field, eddy currents gradually slow down the rate of spin, and spin-oriented satellites must be respun from time to time.

While the orientation of satellites in low orbits may be easier than the orientation of 22,300 mile high satellites, we have too little experience to enable us to evaluate the advantages and hazards of various means of orientation.

IV. DESIGN PROBLEMS IN EQUIPMENT

The power required affects the life profoundly. Life tests show that 5 W travelling-wave tubes have not failed after 4 years, and we believe that carefully designed 2 W tubes, which would be sufficient for carefully-designed satellite communication systems, would have an assured life of 10 years or more. On the other hand, we just don't know how to make long life tubes for hundreds of watts, and even tens of watts pose serious, unsolved problems.

Further, low power is economical from another point of view. Satellite weight is governed by and is nearly proportional to power. A lower power makes it possible to launch more satellites with a given vehicle.

Still another advantage of low power is that it tends to reduce interference.

How can one design and build a system so that a low power aboard the satellite is sufficient? The power required is determined by the noise from the sky, which is a function of frequency, by any thermal noise picked up from the earth, which is very hot compared with the sky, by noise introduced by the receiver, and by the form of modulation used.

In the frequency range between 1 and 10 kMc/s, the noise from the sky corresponds to a temperature of less than 20°K for an antenna pointed more than 7° above the horizon. The noise rises during rain, and this must be taken into account as a possible source of system outage.

A practical maser amplifier adds a noise corresponding to about 10°K, and this noise can be reduced further.

The temperature of the earth is around 300°K. To avoid receiving radiation from the earth, we can use a horn-reflector antenna, such as that used in the ECHO experiment.

The ECHO receiving system had an overall equivalent noise temperature of about 24°K with the antenna pointed at the zenith. Much of this noise was associated with experimental features and could be eliminated. We can expect to attain a noise temperature of 15°K in the near future and below 10°K ultimately.

I have noted that the power required depends on the method of modulation which is used. By using wide-deviation frequency modulation, in which the transmitter frequency is varied in accord with the signal over a frequency range of many times the bandwidth to be transmitted, the signal-to-noise ratio obtained using a given transmitter power is improved. If the frequency is varied up and down from the centre frequency by 10 times the bandwidth, the gain in signal-to-noise ratio is about 100 times. This, however, requires the use of a radio-frequency bandwidth of about 100 Mc/s to transmit a television or other signal having a bandwidth of 5 Mc/s.

Ordinarily, in receiving an f.m. signal, the bandwidth of the receiver must be as wide as the total range of frequencies covered by the received signal—100 Mc/s in the case cited above. But, for the receiver to operate at all in the presence of noise the signal must be around 20 times as powerful as the noise. If the receiver bandwidth is broadened to receive a wide-deviation f.m. signal, the receiver picks up noise, and this minimum ratio of signal to noise may be hard to obtain.

Fortunately, in 1939 J. G. Chaffee described a receiver for f.m. signals with a bandwidth small compared with the range frequencies over which the transmitter sweeps. In this receiver the output is "fed back" to alter the tuning of a comparatively narrow-band receiver, so that the receiver is always tuned to the received signal, however that signal frequency is swept back and forth. The use of Chaffee's f.m.-with-feedback receiver substantially reduces the transmitter power required in a satellite communication system.

We should note that other forms of broad-band modulation, including various pulse modulation systems, could be used to obtain an advantage comparable to that gained with wide-deviation f.m. and an f.m.-with-feedback receiver.

Considering sky noise, the noise of a maser receiver and broad-band modulation, what power do we require in a satellite communication system capable of transmitting a television signal or from 600 to 1000 telephone signals? We will assume a horn-reflector antenna with an aperture 60 ft square. For an unoriented satellite radiating equally in all directions at a height of from 2000 to 5000 miles, the required power is around 2 W. Almost exactly the same power will serve for an oriented 22,300 mile high satellite with a directive antenna that beams a signal over the whole disk of the earth.

Beyond these particular problems, there are many involved considerations of a systems nature. A 24-hr satellite system requires fewer, but somewhat more complicated and heavier, satellites than a low-altitude satellite system. Fewer high satellites would be needed to provide continuous or almost continuous service. However, the failure of 1 satellite in a many-satellite system degrades service scarcely at all.

Further, if some number of satellites assured the availability of at least 1 satellite for a given path, say, 99·9% of the time, 2 satellites would be available for a smaller fraction of the time, 3 for a still smaller fraction and so on. These extra satellites could be used to provide communication over secondary paths on which the greater interruption of service could be tolerated. Of course, any given satellite could be used over one part of the globe at one time and over another part of the globe at another time.

Possible interference between satellite systems and ground microwave systems is an extremely important matter. Several general principles are clear. Anything that makes the powers involved in a satellite system less minimizes interference with ground systems. The low satellite transmitter powers allowed by maser receivers and broad-band modulation will not interfere with ground communication receivers at all. High quality ground antennas with low side lobes and, particularly, horn-reflector antennas, tend to reduce interference.

In attaining many satellite channels at once, one must rely on both frequency separation and on the use of satellites in many parts of the sky which is made possible by highly directive ground antennas. It appears that the use of many satellites will be possible only through the protection against interference afforded by broad-band modulation, so that more actual usable communication will be made available through using a large rather than a small bandwidth in transmitting a given signal.

This problem of interference is very complicated, and there is much to be learned. In the light of our present knowledge it seems best to try to use low powers and broad bands.

The choice among various possible satellite systems must depend on the established state of the art and on a very careful weighing of various considerations. Through the favourable outcome of the ECHO experiment, the capabilities of a passive satellite have been established. It seems to me that, granting an equally favourable outcome following an experimental launching of a simple active satellite, a practical and economical communications system making use of many low-altitude satellites in random orbit need not be many years distant. This is what we are now working toward at the Bell Laboratories.

As time passes, various satellite communications systems will be tried, and I certainly expect that systems involving 24-hr satellites will be built. If at this time I emphasize one general approach to satellite communication it is because it seems to me appropriate to the particular problems and capabilities of these years.

COMMUNICATIONS SATELLITE SYSTEMS SUITABLE FOR COMMONWEALTH TELECOMMUNICATIONS

W. F. HILTON

5, Grange Avenue, Twickenham, Middlesex, England

I. INTRODUCTION

Many of the cities of the Commonwealth are remote from tele-communications trunk routes and the Commonwealth cable system will attempt to rectify this. In the case of cable links, distance will always cost money. The cost of satellite communication is, however, independent of distance, as each call will tie up the same amount of equipment on the ground and in the satellite, with the minor exception of calls involving 2 or more satellites, to the Antipodes. Any comparison of cable and satellite telecommunications systems will therefore be a comparison of systems of very different properties. Commonwealth communications consist of longer links less heavily loaded, in comparison with the transatlantic route, and a possible network is shown in Fig. 1. It is immediately obvious from Fig. 1 that Commonwealth communications are fairly equally distributed north and south of the equator, unlike world communications, which show a distinct peak at about 40°N, as indicated in Fig. 2. Thus the type of solution given in Ref. 1 for economical coverage of the northern hemisphere will not apply in this case, and some form of equatorial orbit will be required. Now the logical steps involved in designing a complete satellite communications system are as follows.

(*a*) Economic survey of the market.
(*b*) Selection of optimum orbit(s) to satisfy demand.
(*c*) Correct sharing of design complexity and cost between ground stations and satellites (system study).

(d) Design of ground station.

(e) Design of satellite.

(f) Design of rocket launcher and launching site.

(g) Development of operating techniques (computers, time multiplex, etc).

This paper will discuss the first 3 of these items.

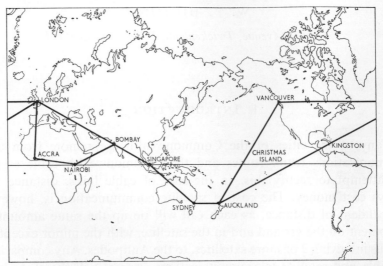

Fig. 1. Map showing Commonwealth cities to be served, and possible relay links.

II. Economic Survey of the Market

A cable will satisfy some of the world's communications demand for all of the time, while one satellite can satisfy a variety of demands for circuits simultaneously or in rotation. In making a survey of the market, we must therefore study the demand from each country, together with the time of day in which this demand will arise. The decision under headings (a) to (g) above are inter-linked to an appreciable extent, so that an intelligent guess as to the rough solution of all problems must be assumed before doing detailed work on one particular aspect. Thus, for example, we must have a rough idea of the plane of the orbit, in order to group the world's telephonic demand into a smaller number of packages, instead of attempting to study a distribution of demand over the whole surface of the habitable globe.

We can, for example, examine the distribution of communications by latitude, as is done in Fig. 2. This is perhaps the correct method for world communications, but not for the British Commonwealth. If, however, we assume that the Commonwealth is best served by an equatorial satellite chain, we package demand between lines of longitude (time zones) and we can represent this demand by the height of a line above the equator.

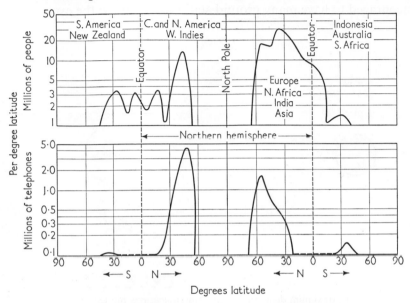

FIG. 2. World distribution of population and telephones by latitude.

Figure 3 shows the result of grouping 30° sectors of longitude on the equator at 2-hourly intervals through the day. The maximum number of channels in use in any direction in space is shown by the height of the line above the earth. For short periods of time satellite orbits stay fixed relative to space axes, so this method of presentation is of value in selecting an orbit. For longer periods of time, of the order of months, the direction of the orbit in space may be changed by perturbations. The direction of the sun will also appear to rotate once per year.

Figure 4 shows the world-wide demand for telecommunications on an hour by hour basis. One fact emerges clearly from Fig. 4. Between the hours of midnight and 0800 GMT the world-wide demand for communications is negligible compared with the rest of the 24 hr. This is true whether we consider satellites or cables, and however we may choose to group the demand into packages.

There is nothing to be gained during these lean hours by having satellites in one place or another—they cannot be used to capacity for intercontinental traffic. This is where the flexibility of a satellite chain will show to advantage over a cable. This surplus capacity can be diverted to more local use during this period. In particular, Australia is at its peak during the midnight to 0800 GMT period and should be encouraged to have many ground stations and to use the excess satellite

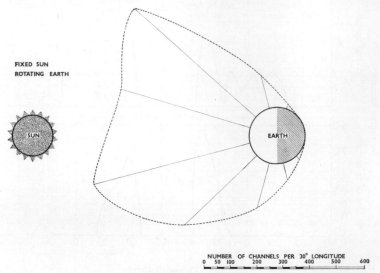

FIXED SUN
ROTATING EARTH

SUN

EARTH

NUMBER OF CHANNELS PER 30° LONGITUDE
0   50 100   200   300   400   500   600

FIG. 3. Maximum demand for telecommunications plotted on space axes.

capacity for relatively short Sydney–Perth–Melbourne–Adelaide–Brisbane links, as well as for top-priority intercontinental traffic.

The greatest intercontinental traffic originates in America and Europe during and just after local business hours, and covers the period 0800 to midnight GMT. The American loading is about double the European and the number of simultaneous international phone calls reaches 880 channels from time to time. So much for the broad economic picture—work is still proceeding at H.S.A. on this never-ending and somewhat non-precise subject. Indeed, market surveys and forecasts can seldom be very accurate. This raises questions when comparisons are made between supply and demand of satellite communications. "Supply" is usually calculated on a digital computer with multi-figure accuracy, while "demand" figures may be considerably in error, and will fluctuate greatly due to the whim of people originating telephone calls.

When it is realized that utilization of the system is demand divided by supply, the need for a careful economic survey becomes apparent, if accuracy is to be achieved. In presenting section XII on annual overall cost, care has been taken to leave the doubtful decision on utilization till the end of the argument, so that the reader can easily insert his own value.

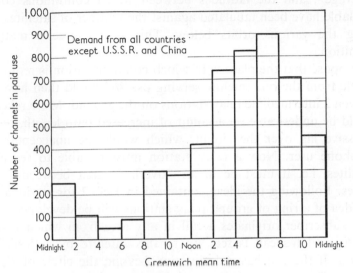

FIG. 4. Diurnal variation in demand for telephone channels.

While mentioning fluctuations of demand, we should remember the Press and radio newsreel uses of telecommunications. A considerable but fixed amount of news material has to be gathered each day, but the place of origin of news fluctuates widely. The flexibility of the satellite system over cables shows to advantage here. Extra channels of communication can be allocated from day to day to those places temporarily in the news, at the expense of others which are quiescent. The constant volume of news demanded by editors to fill their newspapers will result in a constant allocation of satellite capacity.

III. Suitable Orbits and Relay Stations

Having outlined the demand for Commonwealth communications, let us next consider possible orbits to give this type of service. Following Ref. 1, a satellite must rise more than $7\frac{1}{2}°$ above the horizon to give service at a particular point. If it is to give service between 2 points its period of service is determined by the common period of "visibility".

We cannot deal with every conceivable link between all possible combination of 2 places; we can, however, consider the service given by a satellite system to one given point. Suppose another satellite always rises in the west just as one sets in the east. We may call this "continuous coverage", and the latitudes between which continuous coverage is available have been tabulated against the number of satellites required, using the various orbits below. This has a precise mathematical definition.

Suppose that the place with which communication is required is not visible from the one satellite serving us. We would then have to use 1 or even 2 intermediate relay stations on the ground. More relay stations would be undesirable on account of increased path length, resulting in excessive lag over the circuit, which would be unacceptable to the telephone user. Now a relay station must be able to see at least 2 satellites, i.e. it must be in an "overlap" area between 2 coverage circles. Following the ideas contained in Ref. 2 for stationary orbits, the idea of a ring of ground relay stations will be developed.

A subscriber originates a call in any area from which 1 satellite is visible, and his call is sent up to this satellite by a "user" ground station. If the call has a destination beyond the circle of visibility of this satellite, it must be relayed by 1 or more trunk relay stations, each of which must be "working" 2 adjacent satellites. Thus each trunk relay station must always see 2 adjacent satellites. We can have more satellites and less ground stations as we please, within wide limits, and we will now examine the economics of this mathematically.

Let there be n equatorial satellites, each covering a circle of diameter $\theta°$ on the earth, and each overlapping the next circle by $\delta°$ measured on the equator. Since these equatorial relay stations must be able to "see" 2 satellites simultaneously, there will be $360°/\delta°$ relay stations.

The more satellites employed, the fewer ground stations required, and vice versa. If total cost per annum is to be minimized, let us assume each satellite costs p times each ground station per annum, capital costs being amortized in the form of annual costs.

Differentiating and equating to zero, we find

minimum total cost when there are $\sqrt{p}$ ground stations per satellite (1)

Cheapest number of satellites required $= (1 + p^{-\frac{1}{2}})\, 360°/\theta$ (2)

For example a 6-hr circle has $\theta = 120°$ and this leads to Table I following.

TABLE I. Cheapest numbers of ground stations and satellites for 6-hour equatorial circle

Satellite costs 9 ground stations:	4 satellites work 12 equatorial relays
Satellite costs 4 ground stations:	($4\frac{1}{2}$) satellites work 9 equatorial relays
Satellite costs 2·25 ground stations:	5 satellites work ($7\frac{1}{2}$) equatorial relays
Satellites cost 1 ground station:	6 satellites work 6 equatorial relays

Fortunately, large errors in costing only change the balance of satellites and relay stations by a much smaller amount. These relay stations are a necessary evil, and in practice every effort will be made to place relay stations where they can be used by near-by cities for communications, and thus rank as user ground stations.

Various orbits have been considered for Commonwealth communications, and these are discussed below in order of increasing desirability.

IV. 63° SLOT ORBIT, APOGEE AT 63°N

This was described in Ref. 1, and gives a better service to the northern hemisphere at the expense of the southern, and in particular is poor for Australia and New Zealand. It gives excellent avoidance of van Allen belt radiation, can be launched from Woomera, and has no movement of apogee from 63°N latitude. The northern bias is so strong that the London–Vancouver link is as well served as the much shorter but more southerly London New York.

Unlike the orbits described in sections IV–VII below, the mathematically ideal inter-satellite relay stations would not be located on the equator, but above lat. 63°N. Here the basic trunk route for relay stations is less than half the length of the equator, an appreciable economy. It might prove possible to get commercial utilization of all relay stations by adjusting them to cities such as London, Tokyo and Vancouver. Equatorial relay requirements will inevitably lead to ground stations on small Pacific islands with no prospect of local utilization of the available facilities.

V. EQUATORIAL $4\frac{2}{3}$ SUN SEEKING ELLIPSE

This was also described in Ref. 1. It approximates quite well to the demand curve shown in Fig. 3, although England and Canada are

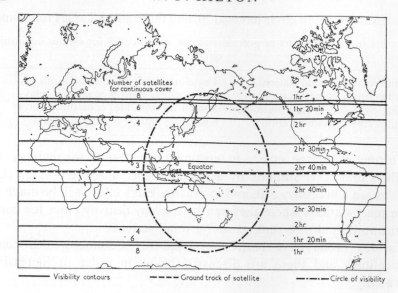

FIG. 5. Visibility contours for 6-hr equatorial circle.

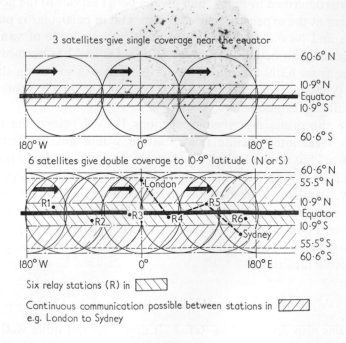

FIG. 6. Ground relay stations must "see" 2 satellites.

fringe areas. It gives no service by night unless it is duplicated on the dark side of the earth. This orbit is bad from the point of view of van Allen avoidance, and cannot easily be launched from Woomera.

VI. EQUATORIAL CIRCLE 6455 MILES ALTITUDE

This 4 times round per day orbit gives 24 hr equatorial coverage, uniform by day and by night. This is undesirable as it does not approximate to Fig. 3. It has excellent van Allen belt avoidance, but cannot be launched from Woomera. It has the poorest weight carrying ability of any of the orbits quoted, for a given launching system.

Being a circle it has one less parameter than an ellipse, and permissible launching dates are no problem at all. It demands a smaller number of equatorial ground stations than most other equatorial orbits.

VII. SUN SEEKING EQUATORIAL ELLIPSE, 5½ TIMES ROUND PER DAY

This ellipse had a day-time apogee of 7600 miles and a night-time perigee of 1200 miles altitude. Apogee will perturb so as to continue to point towards the sun throughout the year. This orbit gives service to a wider area by day, and to a lesser area by night, and may well be used in the future as a communications orbit, particularly when other power sources make van Allen belt avoidance unnecessary.

If we insist on continuous double coverage of the equatorial relay stations by night, it does result in a rather large number of satellites, giving excessive day-time service, unless we have an eye to future demand. The visibility of this orbit is shown plotted on a map in Fig. 7.

VIII. 63° SLOT, 0° APOGEE, 6-HOUR ELLIPSE

All orbits inclined at 63° to the equator have the property that apogee remains over the same latitude. The present orbit is drawn in Fig. 8. If this apogee latitude is made the equator, we have the "poor man's stationary satellite". This orbit can be launched from Woomera, with about 10 times the payload of the stationary 24-hr orbit, which demands an equatorial launching site anyway. For some 4 hr near apogee the angular rotational speed of the earth is fairly well matched by the

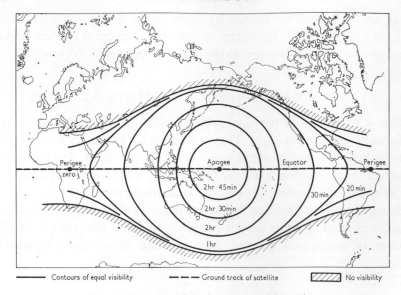

FIG. 7. Coverage of $5\frac{1}{2}$ times sun seeking ellipse.

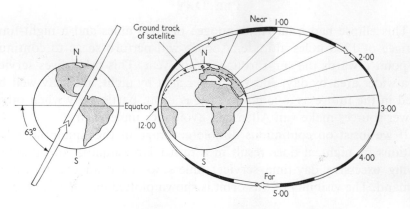

FIG. 8. End and side elevation of 6-hr orbit in 63° slot.

equatorial component of the velocity of the satellite, leaving a residual north to south movement across the equator. This is best seen by plotting the ground track of the satellite on a map, as has been done in Fig. 9.

By plotting circles of visibility at successive time intervals, and noting where they intersect, contours of 1 hr, 2 hr, 3 hr, and 4 hr continuous visibility can be plotted by joining these intersections (Fig. 10).

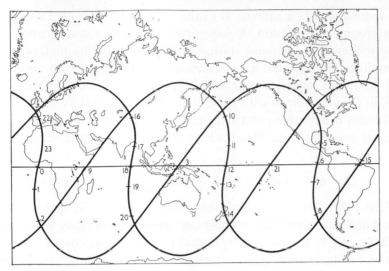

FIG. 9. Ground track of 63°/0°, 6-hr ellipse.

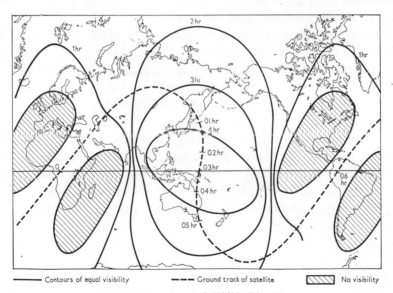

Contours of equal visibility   --- Ground track of satellite   No visibility

10. Visibility times of 63°/0°, 6-hr ellipse.

From Fig. 10 it will be seen that the equatorial belt of 3-hr coverage is about 115° wide so that 8 satellites following a single ground track would give continuous coverage with about 15 near-equatorial ground stations. This is the minimum cost system under the assumption that the annual cost of a satellite is in the region of 3 times the annual cost of a ground relay station. A geographically simpler system using about 12 satellites and 7 ground stations might however be preferable. The 8 satellite system would give continuous coverage between about 45°N and 45°S with intermittent service everywhere outside this area.

This fact, combined with good inner van Allen belt avoidance would make this a desirable orbit if London was not omitted from the continuous coverage area. This leads to the orbit suggested below.

IX. 63° Slot, 20°N Apogee 6-Hour Ellipse

This orbit has all the desirable qualities of No. 7, including inner van Allen belt avoidance, stability of latitude of apogee, which is now at 20°N, etc. In addition, a great deal of continuous coverage is gained in the northern hemisphere at the expense of a smaller loss in the southern hemisphere, thus matching the latitude distribution of Commonwealth cities more accurately. The ground track is shown in Fig. 11 and the visibility times in Fig. 12.

The orbit shares another advantage with orbit No. 3, as it takes advantage of the fact that circles of latitude are in general smaller than the equator, so that a ring of relay stations at 60°N would be exactly half the length of the equator. (It is indeed a pity for polar orbits that a master relay station cannot be established at the North Pole itself!)

The optimum location of the ring of ground relay stations for the present orbit is just below 50°N, and involves (mathematically) 6·9 ground stations. Rounding 6·9 off to give 7 ground relay stations equally spaced in longitude, we find a tolerance in latitude between 45°N and 50° N. A suggested ring of such stations is shown in Fig. 13. Due to the north latitude being more densely populated, only 4 non-profit-making stations need to be established, and these are specially marked in Fig. 13.

This ring of relay ground stations would work 8 satellites following a single ground track, and would yield continuous coverage for all cities from 76°N down to 22°S at the worst, and down to 40°S at the best. This latter variation is due to the slow perturbations of this orbit, and implies that New Zealand would have short gaps in service during some months, but not during other months. Other Commonwealth countries would not be affected, except that very occasionally Australia

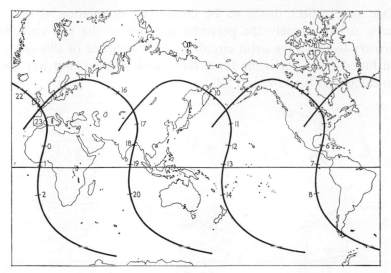

11. Ground track of 63°/20°N, 6-hr ellipse.

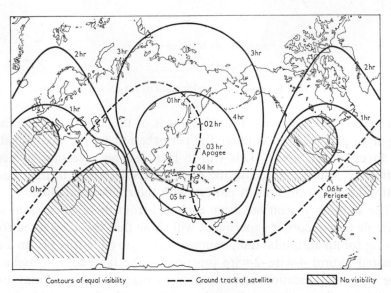

——— Contours of equal visibility   ——— Ground track of satellite   [//////] No visibility

Fig. 12. Visibility of 63°/20°N, 6-hr ellipse.

would have to depend on land line to Brisbane, Queensland, for its inter-continental service. Perhaps an apogee latitude intermediate between the equator and 20°N would improve this coverage without undue loss in the northern hemisphere.

One small effect needs to be considered in relation to all non-circular orbits, namely the perturbations due to the sun and moon, which try to make the orbit circular. The magnitude of this effect has been found to be negligible for orbits 3 and 4 above, but has not yet been calculated for 6, 7, and 8.

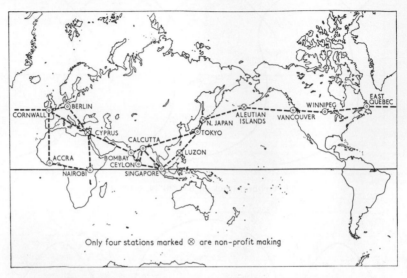

FIG. 13. Chain of ground relay stations for 63°/20°N, ellipse.

However, if important, the advantages of this orbit are such that it might be profitable to counter the perturbation by the occasional use of direct rocket power. This will in fact also be necessary to keep a stationary 24-hr satellite and the other orbits described in their correct station-keeping positions in pattern with their fellow satellites.

The stationary orbit has been excluded from these discussions since a simple tape-recorder experiment will show that the 1200 msec round trip lag for an antipodal link, coupled with echo (side tone) from a mal-adjusted hybrid 4-wire–2-wire junction at the far end will make this type of communication unacceptable to many people. It may prove possible to eliminate the echoes, in which case the lag would be acceptable to more people. The 24-hr orbit would appear to be best suited to television relays, where lag is unimportant. Having discussed at some length the functions of ground stations, we are led to consider the possible compromise between spending money on the ground, and on satellites.

TABLE II

Section of paper	Orbit	Life in years (Radiation hazard)	Relative weight of satellite (Woomera launch)	Relative weight of satellite (Canaveral launch)	Relative weight of satellite (Equatorial launch)	Area of continuous coverage	Area of intermittent coverage	Perturbations
—	300 mile circle	2×10^4	100	100	100	—	—	—
IV	63/63N/4D	22·7	28 (relight)	28* (relight)	28 (relight)	Not yet found	N.P. 70°S	Apogee rotates at 63°N every 2 years
V	0/0/4·2/3D	17·7	16 (4 stages)	16 (4 stages)	33	NOT APPLICABLE	NOT APPLICABLE	Sun seeking
VI	0/0/4D	91	3 (4 stages)	3 (4 stages)	14	8 satellites ±54° 6 satellites ±51°	±60°	None
VII	0/0/5½D	9·9	18 (4 stages)	18 (4 stages)	45	NOT APPLICABLE	NOT APPLICABLE	Sun seeking
VIII	63/0/4D	85	28	28*	28 (relight?)	±45°	N.P.S.P.	Apogee rotates at equator every 2 years
IX	63/20N/4D	29·5	28 (relight?)	28*	28	76°N −22°S	N.P. 40°S	Apogee rotates at 20°N every 2 years

* Involves firing over Cuba, Venezuela and Brazil.
The difference of latitude between Woomera and Canaveral has been ignored.
The difference in the earth's velocity between 0° and 30° latitude (230 ft/sec) has been ignored.

X. Balance of Complexity Between Satellite and Ground Stations

So far we have been considering the number of ground stations, and now we must examine their complexity. This will be done by means of a numerical example. In section II we noted that a world peak usage rate of some 900 channels was called for and, if we assume 18 min use per available channel hr, this implies the provision of $900 \times 60/18 = 3000$ available channels in all.

Should these be split up into many simple light-weight satellites, or should they be concentrated in a few complicated satellites? If we assume orbit 8 with 9 satellites, at any moment 6 are in service and 3 are going through perigee. We therefore require some 500 channels per satellite since $6 \times 500 = 3000$ channels. If, however, we use 80 light-weight satellites, each with only 50 channels, the same saleable product arises, and we can use 80 smaller rockets instead of 9 large ones.

Each ground station will now have to track 10 satellites at once using 10 large dishes plus 1 spare for change-over, instead of 1 plus 1 in the case of the 500 channel satellites. All these numbers will be doubled if transmitter and receiver employ separate dishes. Each pair of dishes together with its associated real estate, which must be sterilized from sources of radio interference or reflection, will cost about £$\frac{1}{2}$ million.

Thus economics appear to favour the 9 heavy satellites as we save £4·5 million per ground station. An additional advantage arises if the satellites have attitude control and station-keeping properties. The associated sensing and control gear will be a constant weight per satellite and thus penalize the 80 lightweight simpler satellites. The advantages of attitude control and station keeping will be considered in a forthcoming paper, Ref. 3.

XI. Availability of Radio Bandwidth

Active satellite relays will enable an adequate number of radio channels to be allocated to serve Commonwealth telecommunications for the rest of this century. Not only does space open up the frequency band from 300 to 10,000 Mc/s for long distance radio communication, but highly directional aerials can be used at these high frequencies, enabling many satellites to use the same frequency allocation simultaneously, discrimination being accomplished by steering the aerials.

In round figures we have 100 times the bandwidth and can work

20 satellites simultaneously on the same frequency from the same ground station by direction finding techniques. This will make some 2000 times the present information bandwidth available by the use of active satellite repeaters.

It would thus appear that we can save precious satellite electrical power by employing wide deviation f.m., and substitute bandwidth for power, while preserving signal/noise ratio.

Ultimately, we shall have cheaper and bigger power sources than solar cells at £0·1 million per kW of d.c. It is hoped that these new power sources will be developed before a shortage of bandwidth arises, by the proposed prodigal use of it in wide deviation f.m.

XII. OVERALL ANNUAL COST

A detailed study of the costing of a Commonwealth communications satellite system is far too complex to include in a paper such as this. It is, however, of interest to give some idea of the sort of figures involved in, for example, a 9-satellite system of the type mentioned in section VIII.

If the satellite life is only a year, and 50% of the firings are successful the 18 BLUE STREAKS and tailored upper stages required, at about 11 each, would cost some £18 million per annum.

Ground stations have also been estimated to cost about £1 million each and may be written off over about 10 years. The cost of a 24-ground station system is then £2·4 million per annum. Maintenance, administration and running costs of 24 ground stations and the computing centre will probably be in the region of £10 million per annum. The total cost of the system, if the above are reasonable, will be close to £30 million per annum.

Nine satellites, each with 720 channels would offer a potential 3396 million channel minutes per annum. If the present intercontinental telephone rate of £1 per minute is charged, 1% utilization must be achieved to break even. This 30 million minutes is to be compared with the present level of some 225 million paid minutes per annum for total world long distance telecommunications. It is of course realized that only part of the existing traffic can be diverted to this new traffic route. High utilization would of course result in very large profit to the operating authority.

The development cost of the rocket, not included above, has been estimated at about £80 million. This could be paid for within a single year if utilization could be increased to about 3·7% of the system potential. In fact a more probable course of action would be to reduce the charges considerably, spreading the development costs over a number of years but stimulating a considerable increase in the total world traffic. Ultimately, with increased demand, the cost of the intercontinental link may be considerably less than the cost of the subscriber-to-dish telephone connection at the two ends.

How can anyone afford not to do this?

XIII. CONCLUSION

The Commonwealth can be served by a minimum of 9 satellites in the orbit described in section VIII. These would have to contain some 720 channels to satisfy the peak demand. Suitable satellites could be launched from Woomera by a rocket using BLUE STREAK as a first stage.

If the go-ahead were given now the satellite system could be in service before the Commonwealth cable in 1967. The latter might then be uneconomic on account of the high rates which will have to be charged.

In conclusion the author would like to thank the many members of the Advanced Projects Group of H.S.A. who have given help with this paper.

REFERENCES

1. DAUNCEY, S. R. AND HILTON, W. F. Communication satellite orbits. XI I. A. F. Congress, Stockholm (1960).
2. BRAY, W. J. The potentialities of artificial earth satellites for radio communication. *J. Inst. Elect. Engrs. London*, **6**, 676 (1960).
3. HILTON, W. F., *et al*. The advantages of attitude stabilization and station keeping in communications satellite orbits. *J. Brit. Inst. Radio Engrs*. September 1961.
4. VARGO, L. G. Orbital Patterns for Satellite systems *J Astr. Sci.*, **VII**, No. 4 (1960).
5. STEWART, B., AND STEWART, P. A. E. Dynamics and Engineering of Satellite Attitude control systems.
6. BUSS, B., AND MILLBURN, J. R. A proposal for an active communication satellite system based on inclined elliptic orbits. *J. Brit. Inst. Radio Engrs*. September 1961.

THE SYNCHRONOUS COMMUNICATIONS SATELLITE

R. P. HAVILAND

*General Electric Company, Missile and Space Vehicle Department,
Philadelphia, Pa., U.S.A.*

I. INTRODUCTION

This paper examines the area of application of synchronous satellites for communications. The term "synchronous satellites" is intended to mean a satellite which is stationary with respect to the earth, that is, one which is in a circular equatorial orbit having a period of 24 hr. Such satellites operate at a height of approximately 19,300 nautical miles.

The approach of the paper is to review certain characteristics of synchronous satellites as compared to lower altitude orbits. These include payload, coverage, time delay and replacement time. The effects of these are then discussed for the relay satellite used to relay messages from one point on the earth's surface to another. The broadcast satellite which sends signals directly to listeners over wide areas is then discussed.

It will be shown that the synchronous satellite appears to be less desirable than lower altitude satellites for the relay service. On the other hand, it will also appear that the synchronous satellite appears to be much superior for the broadcast service.

II. PAYLOAD RELATIONS

The relative energy required to place a given amount of payload in orbit is most easily developed by the "ideal velocity" defined by the equation:

$$V_i = V_{to} + V_{adj} + V_{gravity} + V_{drag} - V_{rotation} \tag{1}$$

The major terms in this equation are the take off velocity $V$to required to ascend to orbital altitude, and the adjust velocity V_{adj} required to remain in the desired orbit. The sum of these terms is plotted in Fig. 1 for a variety of orbital altitudes and for launch sites in and out of the plane of the orbit.

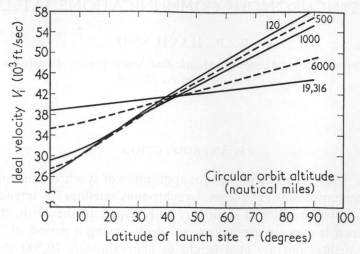

Fig. 1

For example, comparing the synchronous satellite with one operating at an altitude of 6000 miles it is found that the synchronous satellite requires an ideal velocity of 39,100 ft/sec as compared to a velocity of 35,600 for the lower orbit. Higher velocities are required if the launch site is not in the orbit plane. This would occur if the Cape Canaveral launch site is used to establish equatorial orbits. In this case, the synchronous satellite would require velocities of over 40,000 ft/sec while the 6000 mile orbit would require velocities of about 38,000 ft/sec.

Because the velocity required to attain orbit conditions is greater, the synchronous orbit will require a larger booster for a given payload than lower orbits. Alternatively, this orbit will give less payload for a given size booster. This is shown by Fig. 2 which gives the payload *v.* orbital altitude for a hypothetical rocket weighing approximately 300,000 lb at take off and using liquid oxygen and kerosene for the first stages plus hydrogen and oxygen for the last stage. For the equatorial launch site, the synchronous orbit reduces the payload capability for this vehicle from approximately 2500 at the 6000-mile orbit to approximately 1700 lb. Launching from Cape Canaveral causes a further loss, payload being approximately 1000 lb for the synchronous orbit and about 1500 lb for the 6000-mile orbit.

III. Coverage

Because of the greater altitude of the synchronous satellite, it will cover a greater area of the earth's surface than the lower altitude satellites. When comparing the coverage of a system of 10 satellites in

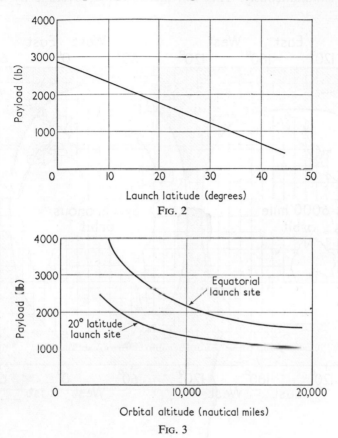

FIG. 2

FIG. 3

the 6000-mile orbit with the 3 satellite synchronous orbit, the coverage is appreciably better for the synchronous orbit, since it reaches 81·5° latitude compared to 68° for the 6000-mile orbit and Fig. 4 demonstrates this improvement.

This difference in coverage, which is shown in Fig. 4 also shows the area which can be reached from a single ground station in the relay service. The relay capability for the synchronous satellite is independent of terminal station location while that for the lower orbit becomes smaller as the station is moved away from the equator. This somewhat greater east-west coverage of the 6000-mile orbit is obtained at the expense of a movable antenna. Also, it should be noted that some station locations with the synchronous orbit will permit the station to see 2 satellites simultaneously. This will increase this coverage by a factor of 2 to 1.

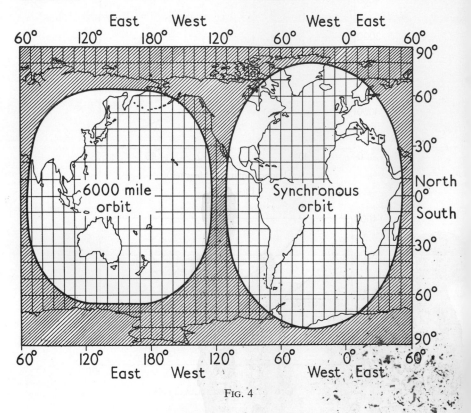

FIG. 4

The required coverage may be investigated by considering the distribution of cities and populations. The 50 largest cities lie between 60°N and 40°S. Only 13 are north of the 40th parallel North. The large concentrations of population are in the temperate zones, between the 20th and 50th parallels. There is a reasonably large and increasing population along the equator and a small population in the sub-arctic areas. Approximately 98% of the population is contained in the band between 60°N and 40°S. Since this band also includes all of the largest cities, it appears that it is the area to be covered by communication satellites. A quick check will show that any equatorial orbit above 5600 nautical miles altitude will give good service.

IV. Time Delay

The time required for a radio signal from the ground to reach another ground station via a satellite relay is shown in Fig. 5, for several orbital

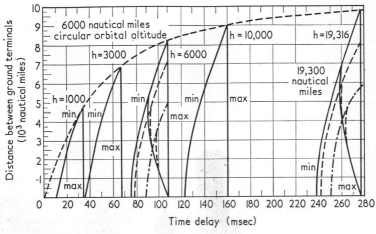

Fig. 5

altitudes. The minimum and maximum delays are shown. For the synchronous system, the time delay is a value which varies from one ground station to another. For other systems, the delay will vary about an average, in the worst case reaching both the minimum and maximum possible delays. However, it is notable that the total variation for a given orbital altitude is relatively small.

This curve also shows the increased coverage made possible by greater orbital altitude. Up to approximately 3000 miles, the coverage increases very nearly with the time delay. Beyond about 6000 miles, the increase in coverage is quite small, while the time delay continues to increase.

To secure greater than line-of-sight coverage, either multiple hops or relaying between satellites must be used. In this case, the time delay is that shown in Fig. 6. Inter-satellite relays at the lower altitudes offer the greatest increase in coverage with the smallest increase in time delay.

V. Replacement Time

It must be expected that catastrophic failure of a satellite can occur, for example, by loss of main power bus in the communication system. In this case, the time to replace the satellite is of importance. The time required during ascent for the 2 orbital altitudes of 6000 and 19,000

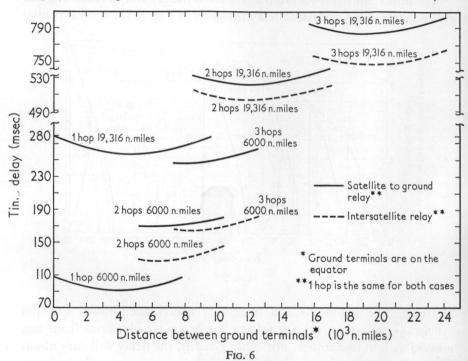

FIG. 6

miles is shown in Fig. 7. It is evident that at the minimum energy condition required for maximum payload, the synchronous satellite requires about 5 hr to reach orbit, almost 3 times as long as for the lower altitude. The time of ascent may be reduced somewhat by using higher burnout velocities, but the gain is relatively small unless quite large velocity is used. This, of course, reduces the payload capacity markedly.

For the synchronous satellite, it appears that a catastrophic failure will definitely give a rather lengthy interruption in service. At lower altitudes, this is not necessarily true, however, when the satellites are installed in the ring system since a replacement automatically comes along. The gap introduced by a failure will reduce the area of continuous coverage or will introduce short interruptions to service but these may be avoided in the relay service by introducing additional up/down hops.

VI. REQUIREMENTS FOR THE RELAY SERVICE

The major requirements for relay satellites are:

> good coverage;
> high quality of service;
> freedom from interruption.

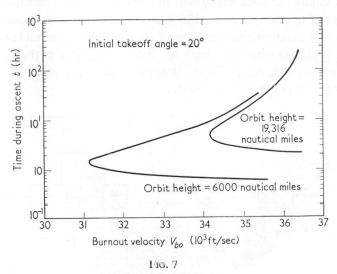

FIG. 7

The coverage has already been discussed. Since most of the international traffic of the relay service flows between highly industrialized areas, and specifically between cities, the terminals to be served lie between about 60°N and 40°S. This indicates that satisfactory coverage can be obtained with any equatorial orbit above about 5600 N miles. (The needed coverage can also be secured with a number of polar satellites at altitudes as low as 3500 miles.)

The effect of loss of a satellite has also been discussed. Other than employing the artifice of positioning 2 satellites at approximately the same point, it appears that loss of the satellite will guarantee interruption of service, if the synchronous orbit is used. This does not need to be true at lower altitudes provided adequate ground relay facilities are installed.

VII. Influence of Orbits on Service Quality

It appears that the only item implicit in the selection of orbits which can affect the quality of service is the matter of time delay and change in time delay. For the telegraphic service, the most important factor is the change in delay. Modern telegraph systems are based on an error correcting query-back. A sudden change in the time delay will cause temporary loss of a character or reception of 2 characteristics simultaneously. Either will cause the transmission to stop until the error is corrected.

This change in time delay will not occur for the synchronous satellite, assuming reasonable station keeping. It is easily avoided for other altitudes by arranging the transfer of traffic from one satellite to another so that it occurs when the delay for the new satellite is the same as the delay for the old. Accordingly, it appears that any orbit will be satisfactory in telegraph type of traffic.

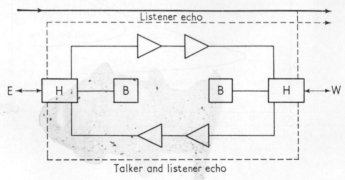

Fig. 8

In the telephone service a change in delay is of minor importance, since delay changes as great as 35 msec are barely perceptible. The magnitude of the time delay in itself may be annoying, however, and when coupled with other factors becomes very serious.

To see this, consider the talker-listener path shown in Fig. 8. Here, as is normal, the telephone is shown as a 2-wire system. A hybrid or 2-wire-4 wire connection is used to separate the signals into 2 paths for long distance transmission. Because these hybrids are not perfect, part of the energy received at the listener hybrid is returned to the sender. Also, part of this echo energy passes through the sender hybrid so that both the talker and the listener perceive echoes. The listener echo is, of course, weaker since it is attentuated twice. Interestingly the talker is the most affected by echo which can cause him to slow down, stutter, or even stop speaking. The magnitude of the effect depends on the echo intensity and on the time delay. Figure 9 shows the C.C.I.F. recommended value of allowed echo intensity as a function of time delay. Two conditions are shown. The upper curve applies to the circuit sketched in Fig. 8. The lower curve is the allowable echo if suppressor circuits are used. The usual suppressor places a short circuit on one line while the other is carrying speech. It is evident that these echo suppressors give a considerable improvement with respect to the allowable echo. However, they do introduce the additional problem of "lockout", which occurs if both persons using the circuit attempt to talk at the same time. In this case, both lines are short circuited and no information is transmitted.

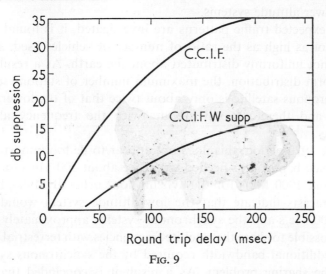

FIG. 9

These problems of time delay, echo and lock-out are quite old. They were first observed in loaded open wire transmission lines which have a very low velocity of propagation. Modern microwave relay circuits are nearly free of the problem. With the satellite system, the problem arises because of the large path length.

A considerable amount of research is currently being conducted on this matter. For example, in their response to FCC Docket 13522, the American Telephone and Telegraph Company reports that "preliminary indications are that a 0·6 sec round trip delay is sufficient to be undesirable and that a 1 sec delay is clearly unsatisfactory".

It is, of course, possible that considerable improvements in echo suppressors will be made. In this case, the higher orbit altitudes could be used. With current equipment, however, it appears that an altitude of about 6000 miles is the largest which would give high quality telephone service.

VIII. EFFECT OF ALTITUDE ON FREQUENCY ALLOCATIONS

On the average, a single synchronous relay satellite must service approximately 3 times as many ground stations as compared to a system of 6000-mile orbits using 10 satellites. Since it appears that the ground stations using a single satellite must be on different frequencies, the occupied bandwidth for the synchronous system will be greater than for the lower altitude systems.

When expected traffic patterns are investigated, it is found that the ratio is not as high as the ratio of number of vehicles used, since the traffic is not uniformly distributed about the earth. As a result of this non-uniform distribution, the maximum number of stations served by the synchronous satellite is only about twice that of the lower latitude satellites and therefore only about twice the frequency allocation is needed.

For the 1970 period, this does not appear to be too important since the occupied bandwidths needed would be about 500 Mc/s for the low altitude and 1000 Mc/s for the synchronous orbit. By 1980, however, traffic forecasts indicate that the low altitude system would require about 1500 Mc/s and the synchronous systems approximately 3000. It is quite possible to share the satellite frequencies with terrestrial services, but the additional bandwidth required by the synchronous system increases the sharing problem. As a result it is concluded that a low-altitude system would be preferred from this viewpoint.

IX. GENERAL CONCLUSION FOR RELAY SATELLITES

From the above brief analysis, it seems reasonable to conclude that the synchronous satellite is not as desirable for the relay of messages between points on the earth's surface as the lower orbits, and, in fact, may not give acceptable service. It definitely occupies a greater percentage of the available frequency spectrum. Service interruptions would be of greater length. For a given amount of service, larger rockets would be needed. Finally, it appears that with current technology, the time delay would be too great to provide satisfactory service. While further developments might give an improvement in service quality, the other factors remain. In view of this, it is concluded that the relay service should be accomplished at lower orbital altitudes with altitudes of approximately 6000 miles being indicated as the most reasonable compromise between coverage and system characteristics.

X. The Synchronous Satellite for Broadcast

When the satellite is used for direct broadcast to listeners, it appears that the dominant factors are the quality of service provided to the listener and the cost to the listener for this service. For example, suppose that 100 million listeners wish to receive satellite broadcasts. This is about equal to the number of TV receivers in the world today. If the cost of special equipment were as little as $10 per listener, the total investment required on the part of the public would be $1 billion. This is a very sizeable sum. To hold this investment to a reasonable level, it appears that the broadcast satellite must operate with current types of receiving equipment. This means that the satellite broadcast service must provide standard signals, at levels typical of current systems and must be able to operate with standard antennas. Movable antennas do not appear to be possible with reasonable installation cost. As a result, it appears that the broadcast satellite system must be based on the synchronous orbit.

To permit estimation of the possibility of the broadcast service a single system will be chosen for study. Since the TV service is one of the most difficult, this is the one used. Later paragraphs will indicate some of the possibilities of establishing a less complex service.

XI. Size of the Television Transmitter

The transmitter power required in the satellite in the broadcast service is most easily calculated by determining the energy per unit area required to give specified grades of service and multiply this by the total coverage of the satellite. Table I shows the r.f. transmitter power required for the 3 U.S. standard grades of TV signals. These are calculated for complete coverage from the stationary satellite. The table also shows the area which can be covered if the transmitter power is arbitrarily limited to 50 kW.

TABLE I. Transmitted power and area coverage of U.S. service grade TV signals

Service grade	Signal level (μV/m)	kW to cover visible zone	Allowable area with 50 kW power (sq miles)
Primary	5010	14,400	0.29×10^6
A	2510	3,600	1.15×10^6
B	224	35	146×10^6

It is evident from this table that a single transmitter can give coverage to the entire visible part of the earth only at the lowest signal value. Where higher grade service is desired, it will be necessary to restrict the coverage area. The "A" grade service with a 50 kW transmitter corresponds quite closely in area to that part of the U.S.A. which lies east of the Mississippi River or to the area of Western Europe. This coverage appears to be ample and has the advantage of limiting the span of coverage to about 2 time zones. This is desirable to permit scheduling programmes during prime viewing hours.

Since the major advantage of satellite broadcasting is the possibility of world-wide coverage, it appears that the satellite should have several transmitters aboard. One or two would be used to give coverage over the visible part of the earth. One or more additional transmitters would be used to give coverage to limited areas of high population density. Early units could use either of these approaches, but it would appear that the wide area coverage would be preferred initially.

XII. ESTIMATE OF TIME SCALE

Fifty kilowatt transmitters do not come in miniaturized form nor is the power demand negligible. Thus the TV service cannot be accomplished with small rockets. An estimate of the possible periods of accomplishment can be obtained from Figs. 10 and 11. These show the current payload weights and power supply capabilities of space vehicles to date and also include a forecast based on constant percentage growth. It is estimated that a 50 kW transmitter weighs 10,000 lb and that the total primary power demand is 400 kW, including an allowance for various audio transmitters. It is estimated that a single 50 kW transmitter can be installed in a satellite in about 7 years. The multiple transmitter system can be expected in the 1970–5 period.

This consideration indicates that the TV broadcast satellite is not

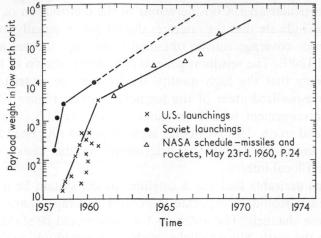

Fig. 10

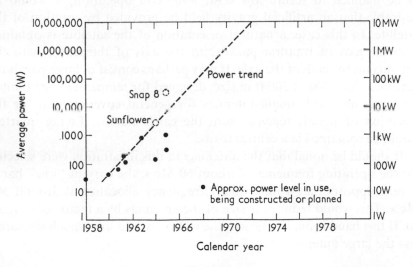

Fig. 11

a pressing matter. However, a period of 10 years is not really very long. This indicates that if we are going to install such a system at the date permitted by technology, studies, discussions and plans should be started shortly.

XIII. Some Considerations for Study

The attractive possibility of world-wide TV service can be accomplished readily by a 3-satellite system, with each satellite positioned to serve a particular area. Considerations of the distribution of land and population indicate that the satellites should not be equally spaced and that superior coverage can be obtained if they are located at 100°W, 30°E, and 120°E. The resultant general coverage is shown in Fig. 12.

It appears that the high quality service may be restricted to the highly industrialized areas of the temperate zones. Figure 13 shows a possible arrangement with 3 transmitters for the U.S., 3 for South America and so on. This particular arrangement is suggested for study. Final arrangements should be determined on the basis of listener density and local interest.

It seems desirable that the 3 satellites in the system be a standard design. Each would incorporate 2 general coverage channels and 6 limited area channels. The antennas for these would be stabilized with respect to the earth. Since vehicles of this size would undoubtedly have to be manned to secure low cost, long-term operation, it would be desirable that an artificial gravity field be provided by rotation of the satellite. In this case, a natural orientation of the satellite is obtained with the axis of rotation parallel to the axis of the earth, with the antennas at the ends of the axle. It may perhaps consist of large parabolic antennas, each 200×300 ft in size, designed for temperate zone limited coverage, and with small antennas for general coverage and for the reception of signals relayed from the earth studios. Living quarters would be contained in a central torus.

It should be noted that the antennas in this illustration were selected for the operating frequency of about 60 Mc/s, the current "low" band. It now appears that a more likely frequency allocation is around 500 Mc/s. This would reduce the size of the antennas by a factor of roughly 10. If this band is finally chosen, the station would look much the same, less the large antennas.

XIV. Simpler Broadcast Services

The fact that the full TV service is extremely difficult is evident from these considerations. This suggests that space broadcasting should start with a simpler service. Probably the best for this would be a high quality frequency modulation broadcast service for voice and music. A

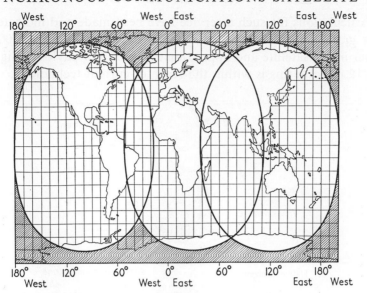

FIG. 12

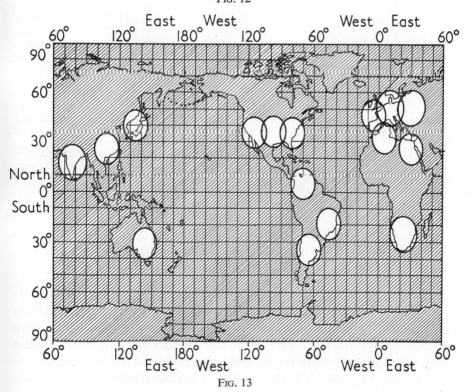

FIG. 13

few calculations for such a service have been made and indicate that a high grade service can be provided with approximately 1500 W of power in the satellite and approximately 500 lb of payload. Figures 11 and 12 indicate this is within the reach of current technology.

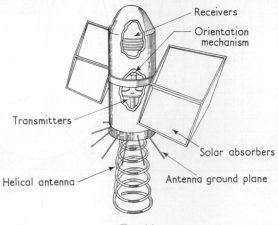

FIG. 14

An artist's concept of a solar power absorber satellite suitable for this service is shown in Fig. 14. This vehicle is designed to keep its vertical axis pointed to the centre of the earth. Rotation about this axis and tilt of the solar absorbers is used to give maximum energy efficiency. A helical antenna is employed to eliminate the effects of vehicle rotation on the signal.

This early satellite could be followed by an early unmanned TV broadcasting satellite, having a single wide area coverage channel. The time period for this would be the 1968–70 period. The large multiple channel manned system could follow in about 10 years, around 1980.

XV. SUMMARY OF FACTORS AFFECTING SPACE BROADCASTING

The possibility of instituting these space broadcast services depends on 4 factors. The first of these is the technical problem of conducting a high quality service from space. It has been indicated in the above discussion that this can be solved immediately for FM broadcasting and within 5 to 10 years for an early single channel TV broadcast.

The second problem is the matter of incompatibility. This is introduced by the differing standards of signals and receivers and by varying frequency allocations. The extent of the problem can be seen from Table II which shows the current TV standards. Obviously, a single standard

TABLE II. Current television standards

Typical country	No. of lines	Field frequency (c/s)	Total bandwidth (Mc/s)	Sound–video separation (Mc/s)	Sound modulation
Austria	625	50	7	5·5	f.m.
Belgium	625	50	7	5·5	a.m.
France	819	50	14	11·15	f.m.
U.K.	405	50	5	3·5	a.m.
U.S.A.	525	60	6	4·5	f.m.
New C.C.I.R.	625	50	8	6·5	f.m.

would be desirable. Two or three standards may be possible, one for each of the zones covered by the 3-satellite system.

Another problem is the matter of securing equitable operation which must take into account the differing habits, customs of various parts of the world as well as the matters of national pride and certainly must reflect differing ideologies and political patterns. To some extent, this problem is eased by considering the 3 satellites separately but it appears that considerable progress must be made in this connection.

The final matter is the problem of language. It would seem that the world-wide system will tend to produce a common language. Many speculations are possible on this but it is suggested that one approach is to "let nature take its course". In this case the world language will probably be determined by the children as they select the programme which they wish to watch.

XVI. Relation of Satellite Broadcasting to Present Methods

The satellite service is inherently a wide area service. It lends itself to covering areas ranging from a major part of a continent to a hemisphere. While smaller areas than this are not impossible technically, it appears that they would be uneconomic as compared to current systems, due to the considerable cost of the booster rockets.

For this reason, space broadcasting should be regarded as an addition to present techniques.

COMMUNICATIONS SATELLITES

G. E. MUELLER
W. B. HEBENSTREIT
E. R. SPANGLER

Space Technology Laboratories, Inc.
Los Angeles, Calif., U.S.A.

I. Introduction

Within the past year several influences have combined to accelerate markedly the activities aimed at establishing communications satellite systems. Principal among these is the increasingly widespread application and acceptance of 2 facts: (1) it is technically feasible to develop a communications satellite system; (2) it is economically profitable to operate such a system. In other words, there is a commercial and military demand for communications satellites, and we have the technology for satisfying these demands.

Technical feasibility rests on eight factors: the rocket hardware and techniques which have been developed largely in the ballistic missile programmes, the improvement of performance and reliability through advanced rocket development programmes, the experience in building, launching, and communicating with satellites in space, a gradually improving knowledge of the space environment and its effects on equipment, electric power supplies which derive their energy either from the sun or from nuclear or chemical fuels, very low-noise receivers, techniques for sensing and controlling vehicle attitude in space, and the ground tracking equipment and high-speed computers needed for accurate ephemeris determination and control.

It has been estimated by the American Telephone and Telegraph Company that about 12,000 equivalent two-way overseas voice channels will be required by 1980. This estimate is remarkable when it is compared to the 345 voice channels between the United States and overseas points which existed in mid-1960. The estimate is based upon a forecast of a thirty-fold increase in overseas telephone conversations in the next 20 years, a substantial increase in international telex and private line teletypewriter traffic, an increase in data transmission, and upon the probable introduction of both closed circuit and broadcast TV. It is, of course, recognized that the actual demand will be sensitive to both the quality of the service and its cost.

The military establishment also has an increasing need for circuits to handle administrative and logistics traffic, which imposes system requirements similar to those of commercial traffic. However, in addition, the military has certain needs which are unique to military operations and which may result in unique system design requirements. For example, the military has a requirement for a type of long-haul communications, including communications with mobile stations such as aircraft, in which the information content is relatively low but the messages are of the utmost importance. This means that the messages must get through during an all-out nuclear attack and in the face of measures (e.g. jamming and physical destruction of the satellite and the ground complexes) by an enemy determined to prevent their getting through. We have, then, on the one hand, certain requirements or demands for communications and, on the other, a varied array of hardware and techniques, both in hand and on the way, which will enable us to satisfy this demand, at least in part.

We should like to address ourselves in this paper to the salient problems in the creation of a communications satellite system and to the trade-offs which are possible in its design; that is, how best do we exploit the available and nearly available hardware and techniques to satisfy the forecasted need for communications circuits?

A communications satellite system includes a large number of complex and complexly interrelated sub-systems. A number of different system configurations are possible (for example, orbit altitude, shape and inclination, operating frequency and bandwidth, number of ground terminals, booster configurations, satellite configuration, etc.), and for each system configuration there is a system design which involves an examination of the interrelations between and the choice of a set of sub-systems to optimize in some sense the resulting system. In many cases these choices and decisions need to be made in advance of complete technical data (e.g. can frequencies be shared with other services and what will be the reliability and cost of rockets in 5 or 10 years from now?) and with sparse and speculative data about customer requirements (e.g. what level of waiting time for circuits will the customer tolerate and what price will he pay for what grade of service?).

More specifically, then, it is the purpose of this paper to discuss the major system design decisions, their interrelationship and their relation to the operational or 'customer' requirements.

II. REQUIREMENTS

Communications links are of three general types: (1) point-to-point trunk in which a great many communications are funnelled to a central point and carried over a single trunk line to another central point from which they fan out to their destinations; (2) point-to-mobile, between a

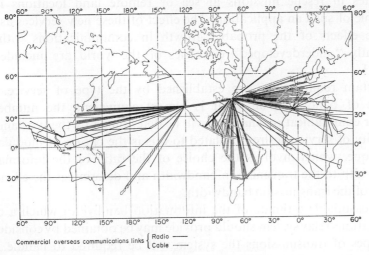

Commercial overseas communications links { Radio ——
{ Cable ········

FIG. 1. Overseas trunk lines used by the United States.

central communication centre and, for example, aircraft in flight or ships at sea; and (3) mobile-to-mobile, such as between the individual aircraft or ships. The great majority of commercial communications are handled over systems of the first type, and many trunk links have had to be established. Figure 1 shows the trunk lines presently used by the United States for overseas communications; the density of traffic handled by these lines in 1958 is shown in Table I.

Communications links are called upon at present to handle three general types of communication: two-way telephone, one-way record message transmissions, such as telegraph or teletype, and one-way television. The world-wide requirements in the first two of these will increase sharply in the years ahead, and possibly two-way television will come into being by 1980. A major new area of growth is the transmission of data of all kinds, for example, in central book-keeping for the world-wide operation of large businesses. Military communications requirements are also growing at a rapid rate. Radio and cable links that are now available for military use are or soon will be saturated by routine administrative and logistic messages. Particularly in the point-to-mobile and mobile-to-mobile types of communications link, the immediate future will bring sharp increases in military traffic.

Finally, we can anticipate a type of traffic which will be brought into being by the existence of communications satellites. The use of these satellites in rescue missions at sea or remote land locations is an example of such an application. The effect of these new services as well as the effects of the probable growth in communications with the presently "underdeveloped" countries of the world are included in Table I.

Certain requirements are established by the type of service. The choice of bandwidth or, what is almost equivalent, the number of similar communication channels will be determined by the type and the quantity of service desired. Related to this is the choice of modulation technique, which involves the choice of the ratio of the information bandwidth to the transmission bandwidth and determines the efficiency of use of the transmission bandwidth.

A reasonable estimate of the information bandwidth which a communications relay system should provide may be obtained by considering the types of transmissions the system will be required to handle. The signals will probably consist of video transmissions, voice transmissions, and teletype transmission, including digital data transmissions. The bandwidth of a video signal, as estimated from the bandwidth requirements for a television signal, is about 5 Mc/s. The bandwidth required for a voice signal may vary, depending upon the quality desired; to achieve toll-quality speech requires a 4-kc/s bandwidth. The bandwidth of typical teletype circuits is about 200 c/s. Similarly, although data-link requirements may vary widely, a value of 20 kc/s can be taken as typical of a relatively high-speed link. Thus, about 100 teletype signals, 5 voice signals, or some appropriate combination of these, can be multiplexed into the bandwidth required for a single data-link transmission.

From the bandwidth requirements for the various types of possible

TABLE I. Present and forecasted overseas communications traffic

U.S. Terminal	Final terminal area	Thousands of telephone messages		Thousands of telegraph messages		Thousands of minutes of telex		Thousands of minutes of Facsimile, etc		No. of television circuits	
		1958	1970	1958	1970	1958	1970	1958	1970	1958	1970
New York	Western Europe	424	4500	4550	5100	875	5000	1600	4800		1
	Central Europe	254	2500	2870	3350	300	2100	600	1800		1
	Eastern Europe	2	20	27	40		50	40	100		
	Scandinavia	42	420	490	580	75	400	160	500		
	Middle East	18	250	228	375		250	35	150		
	Africa	2	50	28	350		225		100		
	Asia	12	180	140	600		325	25	120		
San Francisco	Pacific	260	2600	3053	3500	515	2800	1100	4000		1
	South Pacific	12	250	191	330		400	95	2000		
	Asia	125	1250	1430	3800		3000	140	2000		1
	Indonesia	24	360	270	465		300	25	130		
Miami/New Orleans	Central America	85	820	965	1150	35	850	80	750		
	West Indies	917	5000	10,550	11,500	1200	4900	2000	3250		1
	Northern South America	93	1000	1148	2200		1200	75	800		1
	Southern South America	75	800	935	1660		800	25	500		
Totals		2353	20,000	26,875	35,000	3000	22,600	6000	21,000		6
Equivalent average bandwidth per satellite (3 satellites)		150 kc/s		2 kc/s		200 kc/s		200 kc/s		10 Mc/s	

signals and the forecast of channel requirements in Table I, the band-width requirements for an initial satellite-borne communications system may be estimated.

Although the number of channels required in the satellite varies depending upon the particular pair of terminal points chosen, it is probably economical to build all the satellites alike and to establish the number of channels per satellite and the number of satellites so as to accommodate the maximum forecasted traffic density between any two trunk-line points.

In our study the maximum number of channels required between any two ground points corresponds to an information bandwidth of 15 Mc/s in 1970. This date was chosen since it is believed that the satellites designed today will have a life at least commensurate with their being operable in 1970.

Two-way communications between a pair of ground stations requires a minimum of 4 r.f. channels separated by guard bands: 1 for trans-mission from Station A to the satellite S, 1 from the satellite to Station B, and 2 more for the return path. If 1 or more additional pairs of stations use the same satellite, then additional groups of 4 r.f. channels are required. Each relay pair (e.g. A to S and S to B) will go through a single repeater. Whether or not the other relay pairs go through individual repeaters, or are all frequency multiplexed on a single repeater appears at the present to be a designer's choice.

The choice of modulation used is determined in part by the S/N requirements of the service, in part on the basis of reliability, and in part by the availability of components and power, particularly in the satellite. The basic S/N ratio required for trunk-line telephone com-munications has been established for some years by the Bell System as 45 db.

With respect to reliability of a communications satellite, 2 aspects need to be considered: (1) the error content of the message and (2) the catastrophic failure or gradual degradation of components of the system. Trade-offs are possible in this area, since it is possible to achieve some improvement in (1) by demodulation and remodulation in the satellite, although this may in turn introduce additional components with a resultant decrease in the reliability (2) of the system.

Long-lived components with exceedingly broad-band characteristics have become available in the last few years. In particular, travelling wave tubes with watts of output power and lives of tens of years appear ideally suited to communications satellite requirements. Bandwidths of 100 Mc/s or more are available in these tubes. Satellite power is restricted because: (1) it is expensive in terms of satellite weight; (2) higher powers generally lead to reduced reliability; (3) higher powers increase the likelihood of interference with other services. Here again there is an area of trade-off, since the effective radiated power depends on the antenna gain, but this in turn is dependent on the system philosophy (i.e., illuminate the whole earth or use several antennas to illuminate only the desired area) and on the accuracy of the attitude control system.

The types of modulation which need to be considered include amplitude modulation both single and double sideband, frequency and phase modulation both wide and narrow bandwidth, and various encoding systems. Basically all of these types of modulation except a.m. and narrow band f.m. or p.m. provide a means of more or less efficiently trading output S/N ratios for the ratio of transmission channel bandwidth to information bandwidth. Fortunately for the communications satellite system designer, it is not necessary to determine in advance the particular kind of modulation or even the method of utilizing or dividing the transmission channel bandwidth. The provision of a wide, 100 Mc/s bandwidth in the satellite will permit the user of the satellite to select a modulation scheme which is suitable for his particular service and is flexible enough so that users in different parts of the world may well use completely different modulation techniques on the same satellite.

We conclude, then, that for trunk-line communication, either military or commercial, the channel bandwidth in the satellite should be of the order of 100 Mc/s, which is about the maximum easily obtained with present components, and that the choice of the exact type of modulation should be left to the users.

There is at least one military requirement, however, for which the bandwidth considerations are quite different. This is the requirement for transmissions of low information rate, very urgent messages from a United States command centre, say, to remote highly mobile units (e.g., aircraft or small motorized vehicles) or to remote, hardened bases or weapons posts. For this application, steerable, directive antennas at the ground or aircraft terminal may not be feasible; at most, the antenna will be a fixed, modest sized, resonant structure, such as crossed dipoles.

To obtain maximum system gain using dipole antennas requires lowering the frequency to increase the efficiency of interception of energy. Under these circumstances maximization of system gain dictates the lowest frequency and bandwidth possible consistent with a feasible size for a simple dipole structure. A serious lower limit on the frequency is imposed by the radio blackout effect following detonation of nuclear weapons. Studies indicate that a frequency of a few hundred megacycles is optimum for this requirement.

Because this type of service is characterized by a very small amount of information, but which must get through, the information bandwidths can be small, at most a voice circuit or two or, under certain conditions, even telegraph-type signals or very low rate digital data.

III. ACTIVE VERSUS PASSIVE SATELLITES

A great deal of effort has gone into the study of passive reflectors as satellite relays, because of the attraction of their inherent simplicity and because of the fact that many ground stations could use a single passive reflector simultaneously. These studies have shown that simple passive reflectors do not appear to be as feasible for large capacity trunk-type commercial or military services as the active repeater-type satellite. The following paragraphs compare these two approaches and examine the limitations of each.

The types of passive reflectors considered have included both stabilized and unstabilized reflectors. Stable reflectors are characterized by:

(a) flat plates
(b) direct back-scattering reflectors such as Luneberg lens or corner reflectors, or
(c) shorted antennas, such as an array of resonant elements or a parabolic dish with shorted feeds.

Stabilized reflectors require accurate attitude control, which in turn probably requires electronic systems and thus negates one of the principal advantages (simplicity and reliability) that should be possible with passive reflectors.

The unstabilized or unoriented reflectors studied have included:

(a) a grouping of corner reflectors
(b) a grouping of Luneberg reflectors
(c) metal spheres
(d) a large number of "needles" uniformly distributed throughout an orbit.

Studies of the use of very many small reflectors ("needles") forming a satellite band around the earth seem to rule out this technique for wide-band systems in that a multipath time delay spread as high as 2 msec and doppler frequency spread of 5 kc/s at 8 kMc/s appear inevitable (Ref. 5), in addition to the fact that communication to space through such a band of reflectors would be severely hampered at their resonant frequencies.

The advantage of a passive satellite in the elimination of electronic

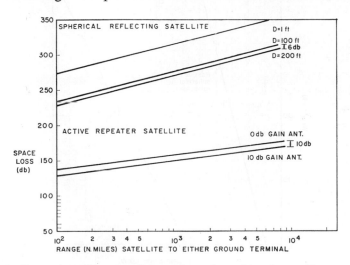

FIG. 2. Comparison of space path losses for active and passive satellite repeaters.

components is balanced by the fact that deformation of the satellite surface itself in time will result in losses in the reflection coefficient of the satellite. The ECHO balloon, for example, although remarkably successful, developed a large scintillation rate as the result of deformation.

The basic problem with the passive satellite with an omnidirectional radiation pattern is its relatively low signal-to-noise ratio for the transmission path lengths required for intercontinental communications. Because it depends upon reflection, its path losses vary as the fourth power of the distance. For an active repeater, on the other hand, because it amplifies the received signal before it reradiates it, path losses are proportional only to the square of the distance. Figure 2 compares the space path losses for both passive and active satellites as a function of the range to the ground station. Doubling the diameter of a spherical reflecting satellite reduces the path loss by only 6 db. The upper curve for the active repeater satellite is for the case of an omnidirectional antenna on the satellite. Clearly the space loss can be reduced by providing a directional antenna on the satellite; for example, a nominal 10 db antenna gain is shown on the figure.

A comparison of the weight requirement for passive and active satellites is shown in Fig. 3. It can be seen that increasing the size of the

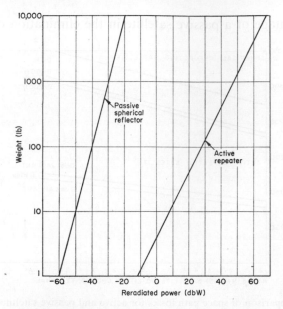

FIG. 3. Weight v. reradiated power at the satellite.

passive satellite to obtain high power gain soon requires a prohibitive cost in weight.

The information that can be transmitted by passive and active repeaters is summarized for a typical set of system constraints in Fig. 4. It is apparent that it is not practical to transmit a voice channel by means of a passive reflector at any altitude higher than about 2000 miles, and even at this altitude it would probably not be possible to achieve the minimum signal-to-noise requirements for toll line communications. The active repeater with omnidirectional antennas is similarly limited by path loss to altitudes less than 4000 miles for toll-quality television. The clear necessity for directional antennas for wideband toll communications in high altitude satellites is shown by the curves for directional antennas.

We conclude that the relatively low information capability possible with passive satellites and the large number of these reflectors which would be needed leads to the choice of active repeaters for commercial communications and for normal military communications. Moreover, as will be shown later, the use of active repeaters permits a significant saving in cost in the creation of a satellite communications system.

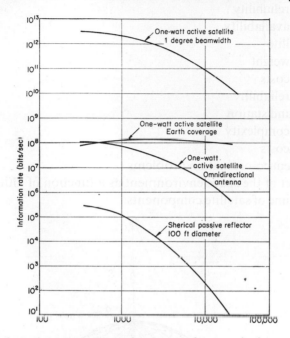

FIG. 4. Information rate *v*. altitude for 4 types of communications satellites.

IV. ORBITS

There is less unanimity today concerning the selection of an orbit than there is concerning the requirements for operating frequency and bandwidth. This is not surprising in view of the fact that the selection of orbits is interwoven with many other system elements and parameters in a relatively complex fashion. For example, the orbit choice either depends upon or is influenced by the following:

(1) the number and location of ground terminals
(2) the number of information channels required between various pairs of ground terminals

(3) the quality of service, especially
   (*a*) the distribution of waiting time for a circuit
   (*b*) time delay for a two-way voice circuit
(4) booster vehicle
   (*a*) payload performance
   (*b*) costs
   (*c*) reliability
   (*d*) availability
(5) satellite
   (*a*) weight
   (*b*) costs
   (*c*) reliability
(6) ground station
   (*a*) complexity
   (*b*) costs
(7) system co-ordination problems
(8) effect of the space environment as a function of altitude upon the
   lifetime of satellite components.

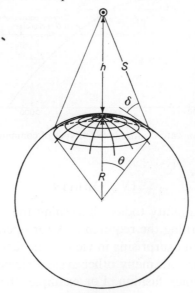

FIG. 5. Global area observed by a satellite.

There are, of course, an infinite number of possible, and perhaps even
feasible, orbital sets. We shall endeavour in this section to develop some
criteria by which we can narrow the field to one or at most a favoured few.

Each satellite will be visible on the earth at any point within a circle whose centre is directly beneath the satellite and whose diameter increases with increasing satellite altitude. As one approaches the perimeter of this circle, however, a greater amount of the earth's atmosphere must be traversed to reach the satellite, a fact which leads at very low

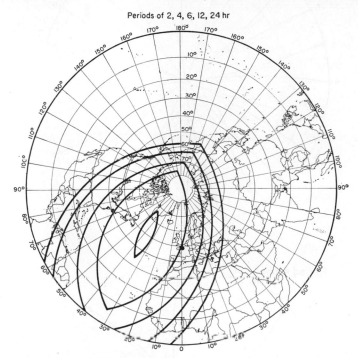

Periods of 2, 4, 6, 12, 24 hr

FIG. 6. Areas of mutual visibility for New York and Paris at various altitudes.

elevation angles to a frequency dependent attenuation. This effect decreases rapidly as the elevation of the source increases above the horizon, and, in fact, can usually be neglected for elevation angles of 10° or more. In addition, a ground antenna aiming far enough down toward the horizon to include a portion of the surface of the earth in its main beam will have an effective noise temperature between that of space and that of the earth. Thus for satisfactory communications the satellite must always be at least a few degrees above the horizon at the point on the earth seeking to use the satellite. This minimum elevation δ can be between 5 and 10° depending on the quality of transmission desired and the type of communications equipment used. In all calculations throughout this paper a δ of 7·5° has been assumed.

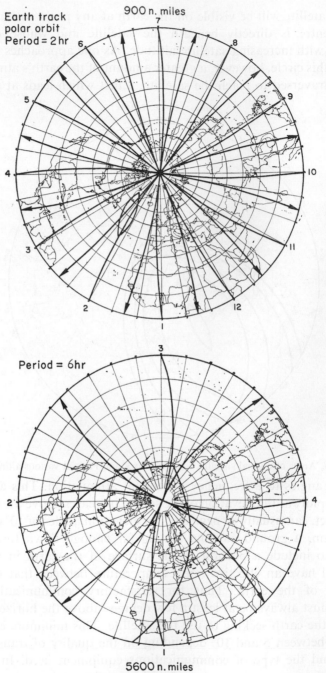

FIG. 7. Suborbital paths for polar satellites in circular orbits at various

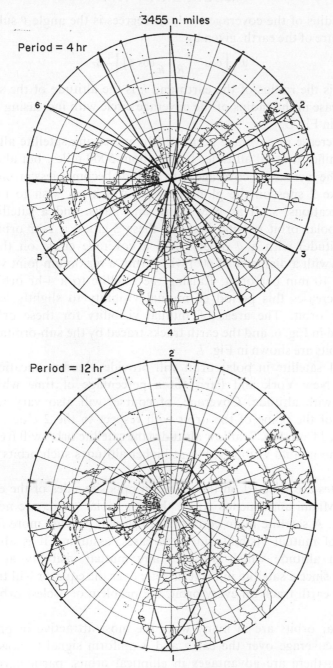

altitudes, showing mutual visibility between New York and Paris.

The radius of the coverage circle in degrees is the angle θ subtended at the centre of the earth, given by

$$\theta = \left[\cos^{-1} \left(\frac{R}{R+h} \cos \delta \right) \right] - \delta$$

where R is the radius of the earth and h is the altitude of the satellite. The increase in the radius of the coverage circle with increasing altitude is shown in Fig. 5.

The increase in coverage permitted by increasing satellite altitude is quite significant at all altitudes below about 5000 miles, but above this altitude the increase in coverage with altitude becomes less pronounced. If we take 2 sites on the globe between which we wish to maintain communications, such as New York and Paris, and a satellite in a circular polar orbit passing between these 2 cities, a 2-hr orbit (1040 miles altitude) with a θ of 30° (and a coverage circle on the earth therefore with a diameter of 60°) will permit a maximum joint visibility of about 10 min per orbit. Raising the satellite to a 4-hr orbit (3962 miles) increases this period of mutual visibility to slightly less than 2 hr per orbit. The areas of mutual visibility for these orbits are illustrated in Fig. 6, and the earth tracks traced by the sub-orbital points of the orbits are shown in Fig. 7.

Thus 1 satellite in polar orbit will provide a communications link between New York and Paris for a percentage of time which will increase with altitude. Coverage, of course, will also vary with the position of the plane of the orbit with respect to the 2 cities. As can be deduced from Fig. 7, at low altitudes the satellite orbit will frequently provide no mutual visibility, but at higher altitudes such orbits are less frequent.

If 2 sites are selected on the globe on opposite sides of the equator, such as Miami and Buenos Aires, then the polar orbits are no longer optimum. A satellite in an equatorial orbit will provide more frequent periods of mutual visibility in these cases, assuming it is above the minimum altitude which provides any mutual visibility at all. In addition, since a satellite in orbit directly over the equator will trace out the same earth track on every orbit, the problem of useless orbits does not arise.

Circular orbits are in many ways the most attractive in providing uniform coverage over the earth and a uniform signal-to-noise ratio. However, there are advantages in elliptical orbits, particularly those inclined at 63·5° which, with the proper choice of periods permit a stationary earth track. The elliptical orbit has the advantage that, with a given booster strength, a greater satellite weight can be placed at the apogee of an elliptical orbit than at the altitude of a circular orbit, when these heights are equal. If the orbit is planned so that apogee occurs during the day in mutual view of densely populated areas and perigee during the night over ocean or desert, an efficient use of the satellite can be obtained. It is possible with a limited number of satellites in elliptical orbits to provide coverage for the daylight hours between most of the communications centres indicated by Fig. 1. The number of satellites in this inclined elliptical orbit needed for continuous communications coverage between 3 pairs of cities is shown in Fig. 8 as a function of

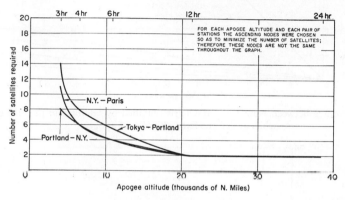

FIG. 8. Number of satellites in inclined elliptical orbit for continuous communications between 3 pairs of cities.

the altitude of apogee, while conversely the coverage obtained by increasing numbers of satellites at fixed apogee altitude is shown in Fig. 9.

However, elliptic orbits have 3 disadvantages. To maintain a constant signal-to-noise ratio from the satellite, it is necessary to vary the antenna gain in elliptical orbits. The perigee altitude for usable elliptical orbits will cause the satellite to pass through the high energy radiation of the inner van Allen belts and, finally, if the perigee is below 200 miles atmospheric drag will result in an unstable ephemeris. We conclude that, except in special cases, the circular orbits are preferable.

To minimize the system complexity the satellite should be above some minimum altitude. As an example, if one wishes to transmit from the west coast of the United States to Japan, a satellite at the midpoint

of the great circle arc connecting these 2 locations must be at an
altitude of at least 900 miles ever to be within the line of sight of both

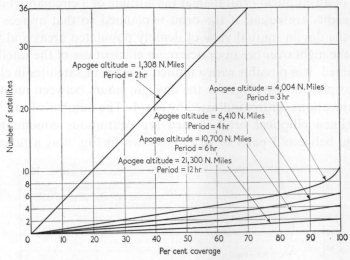

FIG. 9. Coverage between New York and Paris obtained by satellites in inclined elliptical
orbits at various fixed altitudes of apogee.

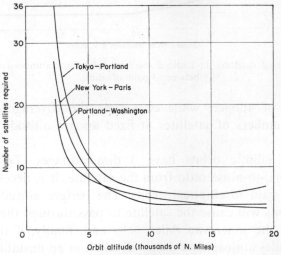

FIG. 10. Number of satellites in circular polar orbits for continuous communications
between 3 pairs of cities.

ground stations. In order to maintain contact with both San Francisco
and Tokyo for an arc of 30° in longitude (which would require a
minimum of 12 satellites), an equatorial satellite must be at an altitude

of more than 7000 miles. Furthermore, it can be shown that for continuous San Francisco–Tokyo communication no fewer than 5 equatorial or 6 polar satellites are required for any altitude, even at infinity, with the single exception of the 24-hr equatorial satellite, in which case only 1 is required. Figure 10 plots the number of satellites needed in circular polar orbits to achieve 100% coverage of 3 pairs of cities.

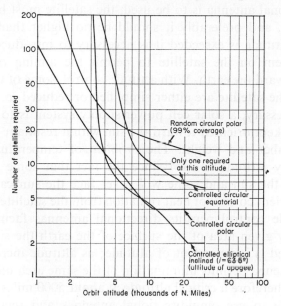

FIG. 11. Number of satellites for continuous coverage between New York and Paris in 4 types of orbits.

The curves of Fig. 10 assume that the satellites in orbit are controlled, both in their initial placement in the orbit and maintenance of that place during the satellite lifetime. Alternatively, at a saving in weight of the "station-keeping" system in the satellite, the satellites can be launched into a specified circular orbit without attempting to control their relative positions in that orbit. To be assured of mutual visibility of 1 satellite between 2 ground sites, a significantly greater number of satellites randomly placed in an orbit needs to be launched, again, however, depending sharply on altitude if it is below 5000 miles. As Fig. 11 shows, 99% coverage between New York and Paris requires 100 satellites randomly located in a circular polar orbit at an altitude of 2000 miles, but only 25 at 5000 miles. To achieve the same coverage between Tokyo and Portland, Oregon, requires a larger number of satellites, and in fact random placing of the number of satellites in orbit to assure communication betwen these 2 cities is at the same time almost equivalent to achieving global coverage.

With omnidirectional antennas on an active repeater satellite, we have seen that it is probably not possible to go above altitudes of 4000 miles and maintain a minimum quality television circuit. Thus if an omnidirectional antenna is to be used, the satellite must be lower than this altitude, and preferably it should be no higher than 2000 miles. Once this altitude is exceeded it is necessary to introduce an attitude control system on the satellite to permit the aiming of directional antennas toward the earth. With attitude control, some of the other sub-systems of the satellite are either simplified or reduced in weight. If, as appears necessary, a solar cell power supply system is to be used, the cells can be kept oriented toward the sun with a resultant major reduction in number and weight. The problem of temperature control is simplified by attitude control since the fact that some surfaces will always face the sun and some will never face the sun can be utilized to effect efficient control of heat radiation from the satellite. In addition, once attitude control permits directional antennas facing the earth, then for any given area of the surface of the earth the signal-to-noise ratio required is independent of altitude; as altitude increases a more directive antenna permits illumination of the same area on the earth.

There is therefore a critical altitude at about 5000 miles below which it is possible to use omnidirectional antennas and relatively simple satellites and above which directional antennas and attitude control are desirable. At the same time, however, there is a sharp decrease in the number of satellites required for altitudes above 5000 miles.

An important consideration in the choice of satellite altitude is the time delay involved in the round trip over the communication link. Studies of the effect of time delay on telephone conversation are now in progress (Ref. 9). It is possible that the round-trip 0·6 sec delay which is introduced by a satellite relay at the distance required for a 24-hr orbit will make that altitude undesirable for commercial telephone relaying unless special send-receive techniques are devised. On the other hand, this time lapse will not impair one-way communications or affect military requirements. In fact, the delay may be useful to military communications in simplifying anti-jamming techniques.

A further consideration in the choice of altitude for a communications satellite is the effect of the space environment on the satellite. Of the

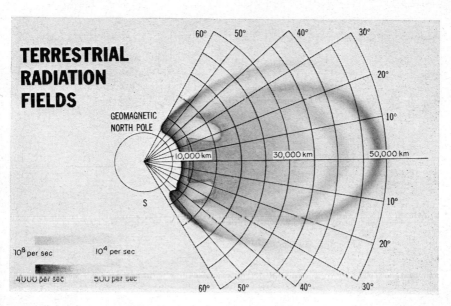

PLATE 1. Terrestrial radiation fields.

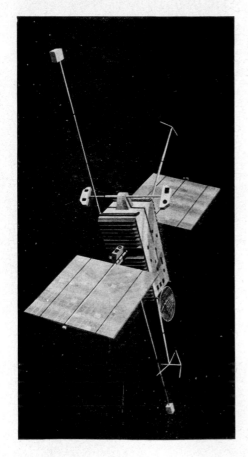

PLATE 2. Three spacecraft exteriors, illustrating their methods of temperature control. *Top left*: orbiting geophysical laboratory; *Top right*: Able 5; *Right*: Pioneer V.

many known characteristics of the space environment, 7 can have deleterious effects: the hard vacuum, micrometeorites, ultra-violet radiation, X-rays, gamma rays, trapped electrons and trapped protons, and of these the last 4 are affected significantly by the choice of altitude. The effects of magnetic fields, infra-red, cosmic rays, neutrons, alpha particles, and higher atomic number particles can be ignored because of their negligible effect or occurrence. (Ref. 3).

Two important effects on solid materials result from the hard vacuum of space which, as shown by Fig. 12, approximates 10^{-12}

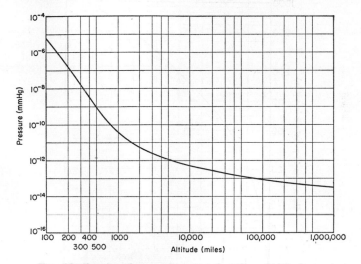

FIG. 12. Density of atmosphere as a function of altitude.

millimetre of mercury in the altitudes from 5000 to 20,000 miles. Sublimation and evaporation are accelerated because molecules leaving the surface of a material are less likely to collide with atmospheric molecules to return them to the surface; and removal of the surface film of gas which covers all material in the sea level atmosphere produces stiction and binding on all surfaces which move in contact with other surfaces. The use of materials with a low vapour pressure, such as plastics, aluminium, beryllium, titanium and gold, can reduce the first problem to negligible proportions and appropriate design to eliminate rubbing contact surfaces can eliminate the second. As can be seen from Fig. 12, once a satellite is a thousand miles or more above the surface of the earth, the variation in atmospheric pressure with altitude is small.

The effect of ultraviolet in the space environment is also to increase

C.S.–F

the rate of sublimation of materials, and can be prevented by the same means, the selection of materials with low sublimation rates.

Measurements of the quantity and energy distribution of micro-meteorites in space to date lead to the curve of Fig. 13, which shows an estimate of the probability of an earth satellite being punctured by a micrometeorite. Puncture by a micrometeorite is possible but, as Fig. 13 indicates, where necessary, as for hermetically sealed batteries, it can be prevented for the anticipated life of the satellite by appropriate skin

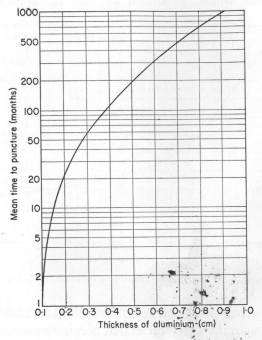

Fig. 13. Estimate of frequency of penetration by micrometeorites in space near earth of an aluminium sphere of 3 metres diameter.

thicknesses. Erosion and roughening of a satellite surface will also result from the impingement of micrometeorites of less than puncture energies, but the undesired effects of erosion on, for example, heat absorption, can be avoided by pre-roughened surfaces or oxide coatings. The flux of micrometeorites has been found in general to decrease with altitude, so that the shielding requirements indicated by Fig. 13 are maximum.

Energy in the X-ray region (1 to 100 Å) and the gamma ray region (0·1 to 1 Å) of the solar spectrum is negligible except during intense solar flares, and the cumulative effect of these flares is not sufficient to cause significant damage to a satellite. The bremstrahlung effect in which some of the energy of particle radiation interacting with a material is reradiated as X-rays and gamma rays can, however, be a factor in damage to satellite materials. Although the energy levels of this secondary radiation are always considerably below the primary, the fact that bremstrahlung radiation is highly penetrating makes shielding difficult, and the effects on semiconductor devices and organic material used in insulation can reduce their operational effectiveness sharply as a function of time. Elimination of the hazards from primary radiation, however, will at the same time eliminate the possibility of damage from X-rays and gamma rays.

The primary cosmic flux consists of particles with energies ranging from a few MeV to 10^9 BeV, with 80 % of the flux made up of protons. Maximum cosmic ray intensity is about 2 particles cm^{-2} sec^{-1}, which is reached at altitudes greater than 15,000 miles over the equator and greater than 10,000 miles over the poles. Maximum ionization in any material due to cosmic rays is 1 mr hr^{-1}, and does not appear to be a significant factor in affecting satellite operation.

A reasonably complete map of the ionizing radiation trapped by the geomagnetic field has been provided by EXPLORER VI (Plate 1). From these data and previous measurements it is clear that there are 2 distinct belts of trapped radiation, the inner belt containing largely high-energy protons, and the outer belt containing largely low-energy electrons.

The principal damage mechanism occurring from particle radiation results from ionization and atom displacement in the affected material, with resultant permanent changes in electrical conductivity and physical characteristics. The low energies of the electrons encountered anywhere in space permit effective shielding with very little addition of weight, but the shielding required by the high energy protons in the inner radiation belt is prohibitive. The incident proton radiation of 20 to 75 MeV can reach counts of 10^4 to 10^6 per square centimetre per second at altitudes of 1000 to 10,000 nautical miles. The highest energy protons, of values 700 MeV and above, are found at altitudes of 2000 to 4000 miles. The damage that this radiation can produce is illustrated by Fig. 14, which compares the short circuit current of silicon solar cells before and after exposure of 740 MeV proton radiation (Ref. 1). An integrated flux of only $7 \cdot 5 \times 10^{10}$ p cm^{-2} at 740 MeV produces a 25%

decrease in solar cell output, and 10^{12} p cm^{-2} reduces the output 50%.

Radiation damage may be prevented by the use of adequate shielding. The shielding required to protect against high-energy protons is shown in Fig. 15. As can be seen, to prevent damage from protons encountered in the inner radiation zone requires at least a 10-in. thickness of the shielding material most effective against protons. From this it appears

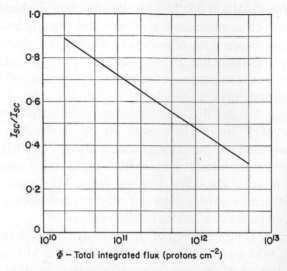

FIG. 14. Short circuit current degradation with 740 MeV protons.

that the only effective deterrent to significant proton damage of communications satellites is to place the satellites at altitudes below or above the inner belt. This would place the satellite below 1000 or above 10,000 miles altitude.

To summarize, we have seen that at an altitude of approximately 5000 miles there is a relatively sharp demarcation in the requirements imposed on the system complexity of a communications satellite on the one hand and in the number of satellites needed for useful coverage on the other. In the region below 5000 miles an active communications satellite cannot long survive unless it is below an altitude of approximately 1000 miles, in which case omnidirectional antennas and relatively simple sub-systems would provide adequate service. Since, for intercontinental visibility, the lowest useful altitude is approximately 2000 miles, we conclude that low altitude active repeaters using solid state electronics are not feasible. In any case, circular orbits of 1000 or 2000 miles will require on the order of hundreds of satellites, to provide significant coverage, and many launches would be required each year to sustain the system.

To bring about significant reduction in the number of satellites launched, we need to go above altitudes of 5000 miles. Here, to assure high-quality communications we need to use active repeater satellites. Once we have chosen active satellites, the space environment forces us

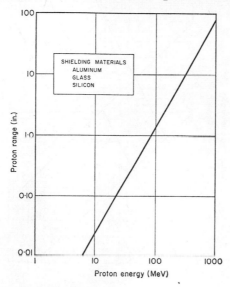

FIG. 15. Shielding thickness required against high energy protons.

to a minimum altitude of 10,000 miles. Since at these altitudes it does not significantly affect launch or satellite complexity or increase cost, there is every advantage in choosing the altitude and orbit requiring the least number of satellites, whether we desire a specific point-to-point coverage or global capacity. The circular equatorial orbit at an altitude of 19,323 nautical miles, unique in its minimum of 1 satellite for point-to-point coverage for almost any pair of major cities in a hemisphere, is a logical choice. If time delay is a real problem for voice communications, then an orbit with a 12-hr period would appear to be the most desirable.

V. CHOICE OF FREQUENCY

One aspect of the choice of frequencies is the characteristics of the transmission medium itself. Satellites in geocentric orbit for extended periods must communicate to earth against the entire range of background radio sky noise, including that of the sun. On the earth, the thermal noise due to the earth's temperature and atmospheric attenuation are affected by frequency.

The problem of selection of an optimum frequency for space com-

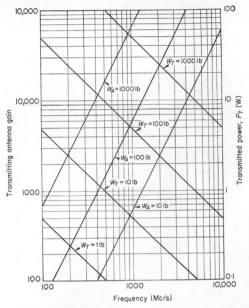

Fig. 16. Airborne antenna gain for 3 antenna weights, W_A, and transmitted power, P_T for 4 transmitter weights both as a function of frequency.

munications has been examined (Ref. 6), with the major constraints illustrated in Fig. 16. For communications satellites with omnidirectional antenna patterns, the optimum frequency for transmitting from the ground to the satellite is about 400 Mc/s, while the return frequency to the ground should, because of galactic noise, be above 1000 Mc/s. For directional antennas and for passive reflectors, the optimum frequency so far as the transmission environment is concerned will lie between 1000 and 10,000 Mc/s.

A second facet in the selection of frequency is the trade-off in weight and performance in the components in the communications sub-system. An empirical relation has been derived from recent experience with high-frequency components which leads to the parametric relations shown in Fig. 16 between the gain, weight, and frequency of the transmitter system, including power supply and temperature control equipment. Essentially these curves are statements of the facts that antenna gain is proportional to the square of the antenna diameter in wavelengths, the weight of an antenna is proportional to its area, and the weight of the power supply increases with frequency since the efficiency of transmitters decreases with increasing frequency. At the present time it appears possible to obtain efficiencies ranging from 50 to 80% for transmitters below 1 kMc/s and from 50 to 10% above 1 kMc/s.

Multiplying the transmitted power by the antenna gain gives the effective radiated power, which is plotted as a function of frequency in Fig. 17. It is apparent that, over the range of frequencies up to 10 kMc/s,

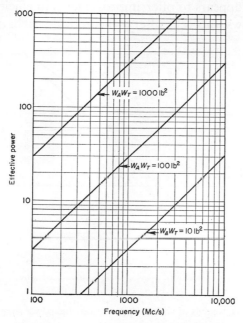

FIG. 17. Effective radiated power as a function of frequency for 3 given weights to be allocated to transmitting antennas and transmitters.

where directive antennas are used, the higher the frequency the greater the effective transmitted power. At frequencies above 10 kMc/s, the requirement for increased mechanical tolerances makes the curves no longer valid. Mechanical tolerances and the increasing atmospheric attenuation found at frequencies above 10 kMc/s place an upper limit of about 10 kMc/s on desirable frequencies for communications satellites.

VI. Choice of Power Supply

The solar cell is available in production quantities for spacecraft requiring electrical power for long durations. Efficiencies of available silicon solar cells range from 7 to 12%, with laboratory versions attaining 14%. Production silicon cell efficiencies should reach 15% in the next few years. Development of new photovoltaic materials, such as gallium arsenide, may allow significant improvements in the efficiency of solar cells, especially in the higher operating temperatures. Maximum theoretical efficiency for the gallium arsenide cell is about 25%, as compared to 20% for the silicon cell. Stacking solar cells of differing energy gaps also appears to offer improvements in efficiency.

As Fig. 18 shows, fuel cells offer a theoretical weight advantage over

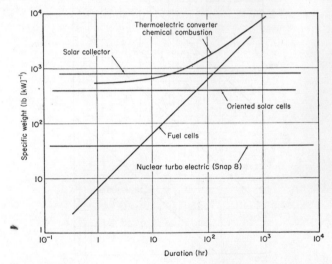

FIG. 18. Specific weights of various types of satellite power supply systems.

solar cells for operation times of less than 100 hr, but no flight type regenerative types of fuel cells have yet been constructed, and it will be several years before they are available. Moreover, for communications satellites, lifetimes of at least 10,000 hr are needed. Figure 18 also compares the present designs for solar collectors as a heat source for a turbo-generator with the oriented solar cell array.

Turbo-generators using a radioactive isotope as the energy source for the working fluid, such as in the SNAP system, are predicted to have a weight advantage over solar cells. The disadvantages of nuclear devices as power sources are the requirement for shielding to protect the rest of the satellite and the danger involved during launching; an explosion at this time could contaminate the launch site if the satellite contained a nuclear power source. Perhaps the major problem, however, is the relatively early state of the development of nuclear power supplies and thus a basic lack of confidence in their ability to provide a reliable power source for long periods of time.

If the solar cells are covered with 0·065 in. plate of glass and backed with 0·025 in. of aluminium, complete shielding from electrons of energy up to 800 keV and protons of energy up to 17 MeV is provided. This shielding is adequate for all orbits except those passing through the lower van Allen belt. It appears that no practical amount of shielding can be provided to assure lifetimes greater than a few days in that zone.

The module construction must be sufficiently massive to yield a paddle with a thermal mass to keep the temperature above −130°C during a 2-hr eclipse.* Ultra-violet and infra-red coatings serve to reduce the absorption of solar energy in those parts of the spectrum where the spectral response of the solar cells is negligible. This decreases the operating temperature of the paddle during the sunlit portion of the orbit and protects the adhesive from ultraviolet radiation.

In conjunction with a solar cell power supply array, provision must be made for storage of power, for which hermetically sealed nickel–cadmium batteries appear to be the available choice. The battery must be designed to supply all electrical power required during eclipses and at those times when the solar cells are not aligned to intercept sufficient solar radiation, which could occur during launch. During normal orbital conditions, sufficient power output must be obtained from the solar cells, then, to recharge the battery as well as to supply the normal satellite load.

The eclipse conditions of low altitude orbits (1000 miles), for which the battery must sustain almost 6000 charge/discharge cycles per year in orbit, provide the worst conditions for battery lifetime. The performance of batteries depends strongly on their design, internal temperature, the depth of discharge during repeated charge/discharge cycles, and the method of charging. To obtain long cyclic life, only a fraction of the battery's rated capacity can be utilized during a single discharge cycle.

We conclude that the best available power supply for communications satellite applications is the solar cell-storage battery system. As

* In the 24-hr equatorial orbit eclipses will be experienced lasting up to 74 min.

more power is required and development experience is gained both solar collectors and nuclear power supplies will become competitive.

VII. Attitude Control

The simplest means of controlling the attitude of a satellite is to spin it about its axis of minimum or maximum inertia; the momentum imparted by the spin will cause the spin axis to remain fixed in inertial space. This method has been successfully used with such first generation spacecraft as PIONEER V.

To apply spin stabilization to a communications satellite at high altitudes, however, does not appear practical. If the spin axis is oriented along the earth's local vertical, it is necessary to provide a constant impulse to cause the satellite to precess 360° per day. With presently possible impulses, such precession would require a prohibitive amount of weight. For a typical angular momentum, 2000 lb ft sec for example, an impulse of 2×10^6 lb sec would be needed. In addition, spin stabilization vertical to the earth would provide at best a solar cell power supply efficiency of 25% of the sun-oriented array.

Orienting the spin axis toward the sun would require a precession of the spin axis at the rate of only one revolution per year, plus that required to compensate for the regression of the nodes of the orbit, and would permit the same solar cell power supply efficiency as an oriented array, but the necessity in this instance to separately orient the satellite antennas would again be expensive in weight. In addition to the weight needed to maintain antenna orientation, to maintain the spin axis orientation to the sun of a 1000 lb satellite for a year would require approximately 5000 lb sec of pneumatic system impulse, the equivalent of 200 lb of nitrogen in an optimum system.

Orienting the spin axis normal to the plane of the satellite's orbit provides a compromise between these two orientations, but again the increase in weight to compensate for reduced antenna gain and power supply efficiency is very large. Solar array efficiency in this orientation in an equatorial orbit would be 28% at best, further reduced for satellite lifetimes of a year to 22% as the result of the variation of the equatorial and ecliptic planes during 1 year. For a polar orbit, maximum efficiency of a solar array would be 25% of a fully sun-oriented array.

As the satellite orbits the earth, it is affected by several torque-producing influences: solar radiation, gravity gradients, magnetic field, micrometeorites, internal rotating parts and atmospheric drag. With the addition of mass expulsion, each of these torques, except that due to micrometeorites, could be used to provide control torques, or these can be treated as disturbance torques and their effects overcome by some other torque-producing system.

The use of solar radiation as a means of obtaining control torques has been examined in some detail; with a large surface, often called a "solar sail", a lever arm can be established between the centre of radiation pressure and the satellite centre of gravity. The required lever arm to obtain appreciable torque from solar radiation is shown in Fig. 19. This

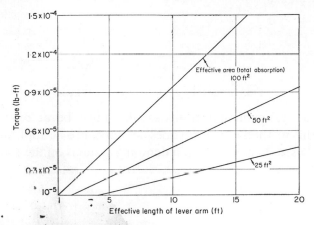

FIG. 19. Torque from solar radiation as a function of area and lever arm.

torque could be controlled either by rotating the satellite or articulating the sail. The requisite size of the sail, its packaging problems during launch, and the need for providing attitude control during eclipse makes this method appear impractical for attitude control of communications satellites.

A satellite which is asymmetrical in its principal moments of inertia will tend to orient with its minimum axis along the earth's local vertical as the result of the gradient of the gravity vector over the length of the body. That is, unless the force of gravity acts along a line passing through the centre of mass, a torque tending to rotate the satellite results. This torque can be employed to stabilize the satellite if its structure is designed with this in mind. The effect of the gradient is important at low altitudes (below 300 miles), and the MIDAS satellite has, in fact, successfully utilized the gravity gradient for earth orientation. The effect diminishes rapidly with altitude, however, and above 1000 miles the torques are extremely difficult to apply effectively. The torque possible from gravity gradient as a function of altitude is plotted in Fig. 20.

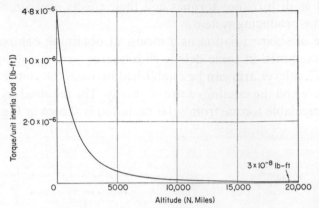

FIG. 20. Gravity gradient torque *v*. altitude.

Since a practical communications satellite must be at relatively high altitudes, this method of control does not appear to be applicable.

Studies based on magnetic spin damping systems on the TIROS 1 and 2

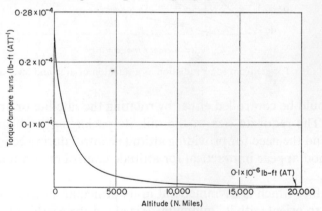

FIG. 21. Interaction of current loop with geomagnetic field.

meteorological satellites show that there is a possibility of utilizing the earth's magnetic field for control system torques. Since the strength of the field at high altitudes is quite low, large magnetic dipoles would have to be carried to achieve sufficient torque; the strength of the torque as a function of altitude is shown in Fig. 21.

It is possible to use the torques created by air molecules in low altitude orbits, 100 miles or below, for attitude control but, again, this source of torque cannot function at high altitudes. In addition, such a method would introduce the need for variable orientation of solar cell arrays with respect to the plane of the orbit.

Attitude control by means of the torque possible from the expulsion of mass (the release under pressure of cold or hot gas, ion propulsion, or plasma propulsion) is also a possibility. Of the mass expulsion methods, cold gas systems are used most extensively, since they are understood most fully and appear to offer the highest reliability. Such systems, however, require more weight than a combination of cold gas and reaction wheels for spacecraft requiring long lifetimes. Figure 22 shows

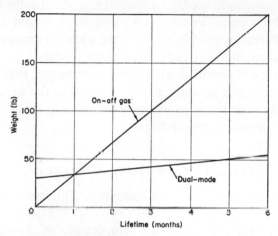

FIG. 22. Weight v. lifetime for attitude control system using reaction wheels and combination of reaction wheels and cold gas.

the trade-off between weight and lifetime of a particular satellite design using cold gas alone and a combination of gas with reaction wheels.

An inertial flywheel driven by a motor provides storage of momentum which can be used either for damping or storing cyclic disturbance torque variations. In combination with a mass expulsion system, the reaction wheel eliminates the need for releasing gas during cyclic torque variations and provides the capability for removing constant disturbance torques. Without a mass expulsion system, the size of an inertial flywheel and its motor drive to handle torques for long operation would be very large. By using a combination of the momentum storage capabilities of reaction wheels and the cold gas nozzle thrust, one can achieve what appears to be the minimum weight for attitude control systems with lifetimes of the order of a year.

One possible embodiment of an attitude control system which combines mass expulsion with reaction wheels is made up of sensors, electronics, and drive mechanisms, in addition to the wheels and pressurized gas. Sun sensors, composed of appropriately shadowed solar cells, are used to obtain error signals representing the satellite or solar array deviation from the sun. Horizon scanners, sensitive to infra-red radiation, are utilized to detect the earth. Electronic circuits provide the necessary amplification, signal conditioning and logic required to actuate the torque sources or drive mechanisms. Torques are provided by pneumatic gas jets and inertial flywheels driven by small servo motors. The momentum storage capability of the flywheel accommodates the cyclic torques arising from disturbances and variation of orbital rates. Nitrogen gas jets provide the moments necessary to null initial rates and counteract steady disturbance torques.

The solar array is provided with servo systems and redundant drive mechanisms for proper positioning with respect to the spacecraft axes. An hermetically sealed unit, using redundant servo motors driving an harmonic gear, positions the solar array.

VIII. TEMPERATURE CONTROL

The temperature of the communications satellite needs to be chosen to provide a benign environment for the components and sub-systems of the satellite. Maintenance of the temperature of the satellite body within the range of 50 to 75°F produces a thermal environment conducive to minimum parts failure rates, an essential feature for a system whose operational lifetime should be measured in years. The efficiency of batteries, solar cells, and many other elements of the satellite vary with temperature; in some cases the additional weight of temperature control devices results in a net saving in the weight of the satellite through optimizing the operating efficiencies of certain satellite components.

Two general approaches to temperature control in space can be taken: passive, in which the satellite surfaces are designed to establish appropriate balances between absorption and radiation of light and heat, and active, in which a change in the internal temperatures activates a control system which reacts to maintain the satellite's internal temperature constant. The passive method, which was used on PIONEER V, as shown in Plate 2, has the obvious advantage of simplicity but the disadvantage that as the surface characteristics of the satellite change with time its ambient temperature will also change. An active system need not be complex; the temperature-sensitive thermal vanes on the ABLE 5 satellite (Plate 2), for example, were simple enough to be as reliable as a passive system. But, however simple in design, an active system will require more weight than a passive one. In practice, of course, the temperature of any surface in space is determined by its exposure to the sun and to other radiating bodies, by its ratio of emittance to absorption, and by its thermal connection to the body of the spacecraft. The design of any space temperature control system, active or passive, is based upon using surfaces for both absorbing and for radiating heat.

If a communications satellite is controlled in attitude with respect to the sun, as appears desirable for other reasons, the problem of temperature control is considerably simplified; the portions of the satellite never facing the sun can be used as effective heat radiators while all the other surfaces can be insulated. A temperature control system which utilizes this fact need consist of only 3 portions: insulation, louvres on the shadowed portion, and appropriately selected surfaces. Since the satellite will always be dissipating power from components within its body, this heat can be used to keep the temperature above some desired minimum, including periods of eclipse. Such a philosophy of temperature control has been adopted for the orbiting geophysical observatory, whose design is shown in Plate 2.

The two sides of the spacecraft normal to the solar cell panel, used as radiating areas, can be constructed of aluminium honeycomb, which has high thermal conductivity, with sufficient facing thickness to distribute the internal power evenly over their surfaces. Major power dissipating components of the attitude control, telemetry, and communications sub-systems can be mounted directly to these sides. The external surfaces can be coated with a material having a value of infrared emittance appropriate to a black body spectrum; such a surface is obtained by utilizing a standard anodizing process.

Control of the radiation can be obtained by temperature sensors and a system of louvres. The louvres are coated with a semi-specular material having a high value of reflectance, such as aluminium sheet or vacuum-deposited aluminium, and can be rotated by means of linkages operated by an actuator such as a bellows filled with a fluid, for example pentane, having a high coefficient of thermal volume expansion. The bellows connect to a tube filled with the working fluid which is attached to the inside of the radiating plates, the length and location being chosen so that an average value of plate temperature is sensed. Temperature variations of the plates are thus converted to changes in louvre position through the volume change of the liquid. This provides a self-regulating, reliable thermal control for the satellite which, when the vehicle is not eclipsed, will permit temperature control within $\pm 2°C$. During the eclipse the temperature depends on the amount of power being dissipated within the satellite; a typical example of expected temperature variation within a satellite is shown in Fig. 23.

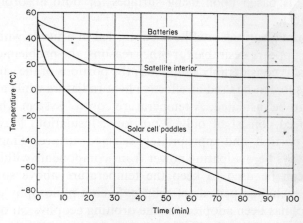

Fig. 23. Typical satellite temperatures as a function of time in eclipse.

The development of "super-insulation" has provided a material particularly applicable to space vehicles. All external surfaces of the satellite body, except for the two sides used as radiators, would in this concept be insulated with this material. The principle of super-insulation is basically one of providing many radiation shields whose surfaces are poor absorbers and emitters of energy at the temperature of the shields. The shields are normally thin sheets of aluminium or mylar with vacuum-deposited aluminium on both sides, and are separated by a material having poor thermal conductivity, such as glass fibre. A typical construction would have 12 to 15 alternate layers of glass fibre and aluminium foil in a thickness of $\frac{1}{4}$ in. The entire assembly is then encased and, for terrestrial applications, a vacuum is produced and maintained in order to eliminate air conduction; in space, the vacuum is automatically maintained.

Portions of the satellite external to the body such as solar cells and antennas are normally best controlled by passive means. The use on the solar cells of quartz or glass plates on which a filter is vacuum deposited can keep their temperature below that which would significantly impair their operation, while during eclipses the lowest temperature reached can be maintained to prevent damage to the cells by providing sufficient heat capacity in the mounting structure.

Trade-off studies comparing the system described above with other possible arrangements have led us to conclude that this combination of passive and active temperature control systems will provide, for a given weight, the most stable operating temperatures for critical satellite components, and is capable of operating reliably in the space environment for very long periods of time.

IX. COMMUNICATIONS SYSTEM

The repeater in a communications satellite requires an amplifier capable of receiving the very low-power signals from the ground, amplifying and off-setting them in frequency, then amplifying again and retransmitting them to the ground. The antenna sub-system may consist either of a simple dipole or, as the result of the trade-offs which we have discussed previously, a directional antenna either permanently pointing at the earth or at a section of the earth depending upon the directivity required by the communications system.

In section II, where the needs of the various communications services were discussed, it was concluded that the optimum active communications system would probably employ a wideband repeater capable of handling a number of different modulation systems; the availability of travelling wave tubes with useful bandwidths of a hundred megacycles and other characteristics of reliability, light weight, and reasonable power outputs make this a feasible system with presently available components.

One possible implementation of such a repeater is shown in block diagram in Fig. 24. Signals are received and passed through a filter to the input of the first t.w.t. amplifier. The filter eliminates feedback of signals, retransmitted by the relay, which will be at a frequency off-set from the received frequency. The first t.w.t. amplifier, operating well below saturation, determines the overall noise figure of the satellite system. The amplified output is then mixed with the output of a local oscillator to provide the necessary frequency shift, and the resulting output is filtered and fed to the final t.w.t. amplifier.

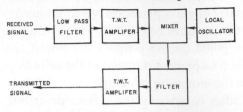

FIG. 24. Simplified block diagram of satellite relay system.

Once a basic choice of a travelling wave tube with a 100 Mc/s bandwidth is made, several second order choices have then to be made. First is a choice between a linear amplifier, where the output amplitude is proportional to the input amplitude, and a nonlinear amplifier, in particular the extreme of a hard limiting amplifier, where the output signal is relatively independent of the input signal level. Studies of the performance of such hard limiting amplifiers have led to the conclusion

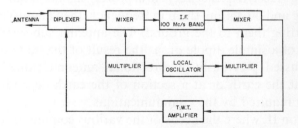

FIG. 25. Block diagram of alternate relay system.

that they are perfectly acceptable for the amplification of multiple channels of f.m. signals. Since the power requirements, the associated system stability requirements, and thus the reliability are higher for this mode of operation, it appears to be a desirable choice. The use of hard limiting amplifiers does result in a restriction on the types of modulation that may be used. Thus frequency modulation, phase modulation, and various forms of pulse code modulation are the most appropriate.

An alternative implementation of the repeater (Fig. 25) makes use of a local oscillator to beat the incoming signal down to an intermediate frequency; an intermediate frequency solid state amplifier, beating back up to the return transmission frequency (off-set from the received frequency); and amplification in a travelling wave tube for transmission to the earth. This system has the advantage of obtaining the major part of the amplification by solid state devices which use very little power and are very reliable. The t.w.t. is used in a region where it also is optimum.

The next significant choice is that of power level. Obviously this depends upon a number of factors, including the ground antenna gain, the airborne antenna gain, the altitude, the background noise, the required output signal-to-noise ratio, the degree of trade-off between bandwidth in the transmission channel and the information channel permissible, the thermal noise levels associated with the receiving amplifiers and other factors.

To summarize the results of a number of trade-off studies, Table II lists characteristics for both the ground-to-satellite and satellite-to-ground links which appear to be a reasonable compromise at the present time. The basic philosophy used in the choice of parameters in this table has been to try to make the satellite repeater as simple and as reliable as possible. Reliability is enhanced by having low power both because of the improved reliability of the transmitting tube itself and because of the simplification and reduced strain on the entire power supply system. Thus the output power of the satellite transmitter has been placed at as low a level as possible consistent with obtaining standard cable signal-to-noise ratios at the output of the ground receiving system. The other characteristics have been adjusted accordingly.

The effective power transmitted from the satellite can be increased by increasing the satellite transmitter power, by increasing transmitter efficiency, or by increasing antenna gain. To increase transmitter power, it is necessary to provide a larger energy source, larger components, and heavier and more complex temperature control mechanisms. All of this results in an almost linear increase in payload weight with increasing power, as seen in Fig. 3. Further we have seen that the life of an amplifier sub-system is a function of power level with a limit of less than 10 W appearing to be desirable at present. Very high-gain satellite antennas can be obtained but at the cost of increased weight and of reduced bandwidth, which limits the coverage and increases the accuracy requirements placed on the attitude control system.

The gain or directivity of a satellite antenna is determined by its aperture. This aperture may be provided by a horn, by a parabolic

TABLE II. Typical characteristics of communication relay links for 12-hr and 24-hr satellites

| | 12-hr orbit | | | | 24-hr orbit | | | |
| | Earth antenna (25° beamwidth) | | 1° Beamwidth antenna | | Earth antenna (20° beamwidth) | | 1° Beamwidth antenna | |
	1 kMc/s	10 kMc/s	1 kMc/s	10 kMc/s	1 kMc/s	10 kMc/s	1 kMc/s	10 kMc/s
Ground to Satellite								
Transmitted power	1 kW	1 kW	1 kW	1 kW	1 kW	1 kW	1 kW	1 kW
Ground antenna gain (60 ft dish)	42 db	62 db	42 db	62 db	42 db	62 db	42 db	62 db
Transmission loss	180 db	200 db	180 db	200 db	185 db	205 db	185 db	205 db
Satellite antenna gain	16 db	16 db	44 db	44 db	18 db	18 db	44 db	44 db
Receiver noise temperature	1200°K	1200°K	1200°K	1200°K	1200°K	1200°K	1200°K	1200°K
Input S/N (20 Mc/s bandwidth)*	29 db	29 db	57 db	57 db	26 db	26 db	52 db	52 db
Satellite to Ground								
Transmitted power	1 W	1 W	1 W	1 W	1 W	1 W	1 W	1 W
Satellite antenna gain	16 db	16 db	44 db	44 db	18 db	18 db	44 db	44 db
Transmission loss	180 db	200 db	180 db	200 db	185 db	205 db	185 db	205 db
Ground antenna gain (60 ft dish)	42 db	62 db	42 db	62 db	42 db	62 db	42 db	62 db
Receiver noise temperature	20°K	20°K	20°K	20°K	20°K	20°K	20°K	20°K
Input S/N (100 Mc/s bandwidth)*	10 db	10 db	38 db	38 db	7 db	7 db	33 db	33 db
Output S/N (5 Mc/s television channel)	35 db	35 db	63 db	63 db	32 db	32 db	58 db	58 db

*Assumes 4 db polarization and other losses.

reflector, or by a linear array of elements. The parabolic antenna is probably the best in terms of bandwidth and weight. The size of the antenna on the satellite will, of course, be a function of the frequency and the desired beamwidth. At 8 kMc/s a parabolic antenna 4·5 in. in diameter will have a beamwidth of 20°, which provides a margin of about 2·5° over the 17·5° subtended by the earth at the altitude of the 24-hr orbit. A 1° beamwidth is obtained at 8 kMc/s with an antenna of 90 in. diameter. At 2 kMc/s an antenna diameter of 18 in. is required to provide the 20° beamwidth.

Finally there is the question of providing telemetry information from the satellite to the ground to describe the operation of the satellite itself. A careful consideration of the complexity involved in introducing telemetry signals into the communications system and particularly the restrictions that then ensue on the types of modulation that can be handled by the repeater has led to the conclusion that a separate telemetry system in the satellite is advantageous and increases the overall reliability of the system. This system is discussed in section X.

In the establishment of a communications satellite system, the need for minimizing weight and complexity in the satellite requires that the design of the ground terminals receive special attention. Improvements in performance obtained by design in the ground station will not require added weight in the satellite. Further, the accessibility of ground equipment makes it feasible and desirable to trade-off complexity and performance on the ground for simplicity in the satellite.

The ground system signal-to-noise can be improved either by increasing the receiving antenna area or by decreasing the effective noise temperature in the input circuits in the receiver. Even with the best of maser pre-amplifiers, the input noise temperature cannot be reduced much below 20°K because of the antenna radiation characteristics. Therefore, no further improvement in the effective noise temperature of the receiver appears practical.

The diameter of the receiving antenna is related to the other system parameters by

$$D = 4R \left(\frac{l\, k\, T\, B\, n\, S/N}{P_T G_T \epsilon} \right)^{1/2}$$

where D = the diameter of the ground receiving antenna (ft), $kT = 1\cdot3 \times 10^{-23} \times 300 = 4 \times 10^{-21}$ J, R = the range to the vehicle (ft), n = the noise figure of the receiver, S/N = the required signal-to-noise ratio, P_T = the power transmitted from the vehicle (W), G_T = the gain of the transmitting antenna in the vehicle, $B =$ the bandwidth (c/s), $l =$ the polarization, cable, etc. loss, $\epsilon =$ the antenna efficiency.

Using the following values of the parameters for the communications satellite system, $R = 1.16 \times 10^8$ ft, $n = 1.07$ or 0.7 db, $S/N = 50$ db, $P_T = 1$ W, $G_T = 18$ db, $B = 100$ Mc/s, $l = 3$ db, $\epsilon = 0.5$ indicates that a ground antenna diameter of approximately 30 ft is the minimum necessary. In the actual design, the use of parametric pre-amplifiers in the ground receiver can increase the value of n to approximately 2 db. With these values an effective antenna diameter of 50 ft can provide adequate reception. To provide flexibility and additional operating margins it seems desirable to provide antennas with diameters of 60 ft at the ground stations.

The size of the antenna which can be used at the ground station is limited by the accuracy with which the satellite can be tracked and by the dimensional accuracy in terms of wavelengths to which the antenna can be constructed and maintained. Maintenance of the necessary tolerances for use at 10 kMc/s is the limit of our present capabilities for a 60 ft parabolic antenna, and this will provide a beamwidth of $0.1°$. This beamwidth is compatible with the tracking accuracy obtained from the t.t.c. and will provide a gain of about 63 db.

To achieve the minimum possible ground receiver noise temperatures it is necessary to use a special horn-reflector antenna or its equivalent such as that developed by the Bell Telephone Laboratories for its microwave relays. With this antenna and a maser pre-amplifier it is possible to achieve a noise level of about $20°$K at angles greater than $10°$ above the horizon. This is the noise temperature used in the calculation of Table II. Obviously there is a trade-off between the cost of these special horn antennas and the much simpler parabolas as there is between masers and parametric amplifiers. It appears that the basic needs of the communications satellite system can be adequately met by steerable parabolic antennas of 60 ft diameter and with a parametric pre-amplifier.

On the link from the ground to the satellite it is reasonable to use the same effective antenna aperture. It is feasible to use diplexers to separate the ground transmitted frequency from the received frequency and this is probably economically more desirable than using a separate ground transmitting antenna. The ground transmitter power is calculated from

$$P = \frac{16R^2 \, l \, k \, T \, B \, n \, S/N}{D^2 \, G_T \epsilon}$$

and for a 40 db signal-to-noise ratio at the satellite input, a 2 to 1000 W amplifier is required depending on the satellite antenna gain. In order to provide operating margin a 1000 W amplifier is probably desirable in any case.

X. Telemetry, Tracking and Command

To gather information on the location and condition of the communications satellite, and to control its operation, a separate system is required, with components both in the satellite and on the ground. As suggested in section IX, it would undesirably complicate the communications system itself to utilize this r.f. link for telemetry, tracking and command, but it is feasible to combine these 3 functions into a single integrated system, as was done, for example, on EXPLORER VI and PIONEER V. The integration is possible in that the ground-to-satellite r.f. link necessary for transmitting commands can simultaneously be used as the signal which, transponded and returned, permits ground tracking of the satellite. Similarly, this r.f. link to the ground for tracking purposes can also be modulated with telemetry information. The integration is desirable in that multiple applications of the receiver, transmitter and antenna system in the satellite permit significant savings in weight and complexity, and with no impairment of reliability. Since failure of any one of the 3 elements of telemetry, tracking or command in an active repeater using position and attitude control makes the operation of the other two useless, there is no increase in overall satellite reliability by separately instrumenting the functions.

Gathering of information from the satellite falls into 2 categories: tracking information by means of which the position and ephemeris of the satellite can be determined and telemetry information on the condition and operation of the satellite and the communications repeater. On the basis of these two types of information, the ground tracking and data handling system computes the necessary ground antenna steering data and determines the proper commands to be transmitted to the satellite. Normally a single central data reduction and analysis centre is desirable for computing satellite position and condition, using high-speed computers, and to facilitate the operation of this centre, digital techniques for both telemetry and commands are preferable.

For a satellite of the type required in a 24-hr orbit, approximately 60 different commands to the satellite must be provided for, to adjust orbit position and satellite attitude, control transmitters, operate the communications equipment, and other functions. A similar number of satellite functions need to be monitored, temperatures, voltages, currents, tachometer readings in the attitude control system, gas pressures, and so on. Tracking is obtained by doppler determinations of range rate.

The slowly varying nature of most of the telemetered information permits a relatively narrow information bandwidth, on the order of 300

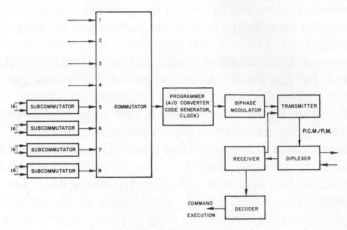

FIG. 26. Block diagram of t.t.c. system.

c/s for adequate telemetry, an individual measurement being telemetered at a sampling rate of, for example, twice per sec, 10 sec per sample, or even more slowly depending on the critical nature of the measurement or its probable rate of change.

Of the many possible methods of encoding telemetry on to the satellite-to-ground transmission, p.c.m. or p.a.m. are the simplest and hence most accurate, in view of the requirement for relatively slow sampling of functions in a narrow information bandwidth. Since p.c.m. offers a substantial superiority in signal-to-noise ratio, it appears to be the logical choice for a pulse coding technique. Using an 8 bit word, 2 bits of which are for identification, a p.c.m. rate of 256 bits sec^{-1} will adequately carry the telemetry information.

A possible arrangement of the sampling sequence is suggested by Fig. 26 in which 3 measurements are transmitted on each data frame, but 64 are subcommutated, to be transmitted every 16th frame. As Fig. 26 shows, the p.c.m. signal frequency biphase modulates the carrier.

Assuming a 400 Mc/s satellite-to-ground t.t.c. carrier frequency, the following system performance can be anticipated.

Satellite transmitted power (1 W)   30 dbm
Satellite antenna gain   0
Ground antenna gain   35
          ──
                    +65 dbm

Transmitter line loss   —1 dbm
Space loss   —178
Maximum fading ($\delta = 5°$)   —6
Receiver line loss   —2
Polarization and scan losses   —6
          ──

                    —193 dbm
          ────
                    —128 dbm

Experience has shown that tracking and telemetry receivers with sensitivities on the order of —150 dbm are readily available, and therefore

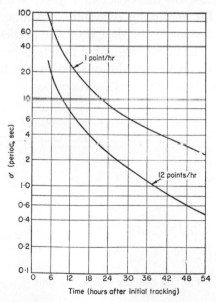

FIG. 27. Increased tracking accuracy with time (satellite in 20,000 mile orbit).

the satellite-to-ground link can operate with an omnidirectional antenna. It appears desirable to utilize an omnidirectional antenna for the t.t.c.

so as to provide commands and telemetry even if the satellite should be tumbling.

Measurements of the doppler shift of the r.f. link of the t.t.c. system will permit continuous measurement of the satellite range rate. The satellite transponder off-sets the frequency transmitted from the ground by a rational fraction; instrumentation at the ground station, by comparing the transponded satellite-to-ground frequency with the ground-to-satellite frequency, permits very accurate round-trip doppler measurements. These measurements permit a determination of the velocity of the satellite within 1 ft sec⁻¹, with 1 sec smoothing. Angular measurements by the ground antenna, averaged for 1 sec, will permit accuracies within 1 mrad. Tracking a satellite in 24-hr orbit over an extended period of time will permit the accuracies indicated in Fig. 27.

XI. Reliability

If a communications system using satellites is not to become prohibitively expensive, the satellite must be an extremely reliable system, with a lifetime at least on the order of 1 year. We have applied the classical procedure for predicting reliability as a function of time for a complex electronic system to the design of a hypothetical communications satellite such as has been suggested in this paper and have found on the basis of conservative calculations that it can have a mean time before failure of more than 1·5 years.

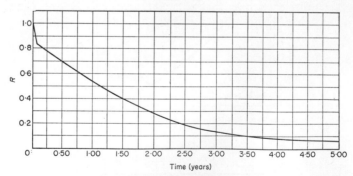

Fig. 28. Satellite reliability *v.* time.

The curve of satellite reliability as a function of time is shown in Fig. 28. The variation from a Poisson distribution results from the utilization of redundancy. The truncation at 5 years results from the fact that the satellite carries a 5-year attitude-control gas supply. Calculations based on this curve indicate the satellite mean time before failure of 1·5 years.

A degradation factor of 0·33 has been applied to the failure rates of electronic and electromechanical parts in an attempt to compensate for the uncertainties which still exist concerning the space environment. It is quite possible that the effects of environmental degradation have been over-estimated and that a factor of 0·5 or 0·7 would be more applicable. Use of these factors would increase the 1-year reliability potential of the satellite to 0·6 and 0·7 respectively, and reduce the anticipated number of launches required accordingly.

In this study catastrophic failure of the satellite was assumed when either of 2 conditions applied: (1) inability of the communication sub-system to transmit and receive at sufficient power levels and (2) satellite drift from the required position by 1·5°.

Intelligent application of the means possible for increasing reliability was assumed, properly utilized redundancy, design margins, temperature control, packaging, and testing. It was calculated that an active repeater satellite with active temperature control and attitude control would require 500 in-line active elements, providing a mean time before failure of 4000 hr. Careful integration of 250 redundant active elements increases the mean time before failure to 12,000 hr. In addition, it is assumed that the maximum stress applied to any part is 25% of rating and that the temperature in the satellite body does not rise above 86°F at any point except the output stage of the power amplifier.

It is probable that major improvements in the component reliability will be achieved in the next few years. Based upon early results of this programme it is possible to predict that the lifetime of communications satellites can be extended to more than 10 years.

XII. Cost

In establishing and maintaining a communications satellite system the major costs will be distributed among 3 elements, the satellites themselves, the launches and the ground stations; the costs of these elements will depend of course on the quantity and complexity involved in each case. The satellite cost will be minimum for small, passive reflectors and maximum for active repeaters with directional antennas and attitude controls. The launch cost will vary as the result of the use of relatively small rockets and simple launch procedures to place passive satellites in low altitude orbits to the large boosters and more complicated procedures needed to place active repeaters at high altitudes. The power requirements and antenna sizes of the ground stations will increase with the altitude of the satellites, but at the same time the angular tracking rate of the ground antenna will decrease as the altitude of the satellite increases.

Combining the cost of the satellite, booster rockets, and launch procedures into an approximate cost per launch provides a range of $500,000 to $5 million for repeater satellites which do not maintain themselves in position in orbit, the cost increasing with the altitude of the satellites (Ref. 2). Similar approximations provide a range of $5 million to $9 million per launch for satellites in controlled orbits. The cost of each ground station is estimated at $2·5 million, regardless of satellite altitude.

Conservative estimates indicate that it is reasonable to expect a mean time to failure of a repeater satellite of 1·5 years and a successful launch probability of 0·75 by late 1963. The cost of establishing a communications satellite system can be approximated by the formula

$$C_E = C_g N_g + \frac{N_s}{p\,n}(C_B + nC_s).$$

The expected annual replacement cost is given by

$$C_M = \frac{N_s}{p\,n\,t}(C_B + nC_s),$$

where $N_g = $ the number of ground stations, $N_s = $ the number of operating satellites required in the orbital set, $p = $ probability of a successful launch, $t = $ average lifetime of a satellite, $n = $ number of satellites launched per booster, $C_g = $ cost of each ground station, $C_B = $ cost of booster, launching operation, and range time, $C_s = $ cost of satellite vehicle. The estimates of cost obtained by these means are useful principally for comparisons between systems and not for careful estimates of the design of a given system, since such factors as amortization of development costs and maintenance of an adequate inventory of spare equipment are not included.

The estimated costs for establishing and maintaining a communications satellite system for a particular (New York–Paris) point-to-point trunk line system and for global coverage are presented in Figs. 29–33, considering the costs both for randomly located and controlled satellites.

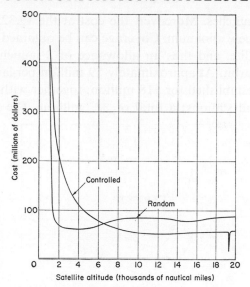

FIG. 29. Cost of establishing communications satellites for continuous coverage, New York to Paris.

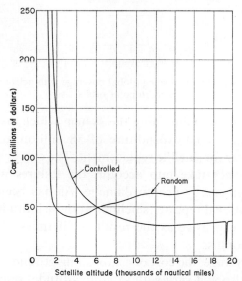

FIG. 30. Annual cost of maintaining communications satellites for continuous coverage, New York to Paris.

Figure 29 shows that above 6000 or 7000 miles altitude it is less expensive to establish and maintain coverage between New York and Paris by means of controlled satellites, despite the increased cost of the individual satellites and launches. Moreover, the cost at the 19,323 nautical-mile altitude is uniquely minimum. Coverage can be obtained at this altitude by only 1 satellite, and thus an allowance of 2 launches to establish the satellite in orbit. At approximately $9 million per launch, then, the system can be established for $18 million, together with the $5 million for the 2 ground stations: a total of $23 million. The same rationale

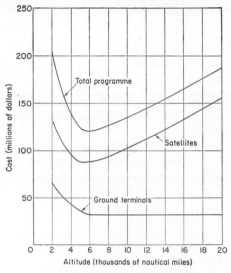

FIG. 31. Estimated costs for establishing global coverage with active repeater satellites in random polar orbits.

applies to the cost of maintaining the system which, as Fig. 30 shows, would require an annual expenditure of $8 million, using a figure of 0·9 launch per year for satellite replacement. The curves of Fig. 30 below 10,000 miles do not take into account the failures which would be produced by high-energy proton damage. As has been shown, maintenance of operational repeater satellites at these altitudes would in fact be prohibitively expensive as the result of lifetimes of days or weeks rather than 18 months.

For global coverage, the cost of establishing ground stations becomes a factor which also will vary with satellite altitude, since more stations will be needed to assure point-to-point communications throughout the world when satellites are in lower orbits. As Fig. 31 shows, the minimum cost for establishing global coverage with satellites randomly located in orbit will occur when circular polar orbits of about 6000 miles altitude are used, including a cost of approximately $32 million for ground stations and $93 million for satellites. The minimum for controlled satellites (Fig. 32) occurs again at the 24-hr synchronous altitude, an estimated total of $70 million, including the 4 launches probably required to place 3 satellites in synchronous orbit. An annual maintenance cost of $18 million to cover 1·96 launches is required for

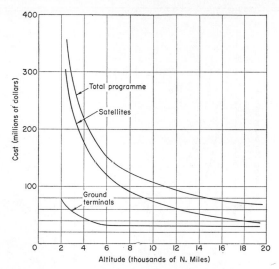

FIG. 32. Estimated costs for establishing global coverage with repeater satellites in controlled orbits.

this system. Again, the curves of Fig. 33 below 10,000 miles need to be multiplied by a proton-damage factor before they can be considered realistic.

If we eliminate all orbits below 10,000 miles, we have in effect eliminated all but 3, namely: (1) 12-hr circular polar orbits, (2) 12-hr circular equatorial, (3) 24-hr equatorial.

On the basis of cost alone we could eliminate the 12-hr orbit sets and conclude that the optimum would be the 24-hr equatorial system. However, the two-way signal time delay in the 24-hr system may preclude its use for commercial telephone circuits. The question of whether or not this delay will constitute a serious problem for two-way voice communication is yet to be resolved. A limited number of tests have been made which simulate both the delay and the lock-out occasioned by the action of the echo suppressors. Unfortunately various studies have given conflicting results and so we do not have sufficient criteria for deciding between a 12-hr polar set and a 24-hr equatorial set. The former may be between 2 or 3 times more expensive to maintain than the latter, but a carrier may be pleased to pay the price.

Submarine cables are the chief competitor of communications satellites for transoceanic communications, and a comparison of the costs for cables and for satellites for similar service shows a marked

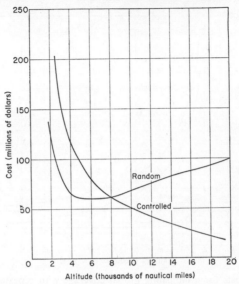

Fig. 33. Estimated annual cost of maintaining global coverage communications satellites.

difference in favour of the satellite. Using as a basis the new type of cable which AT and T plans to lay in 1963, and the 24-hr system described in this paper, it can be found that the annual cost of a 3000 mile transoceanic cable link would be about $27,000 per year per voice channel, compared to $10,000 per year per channel with the satellite.* Overland microwave relays as now used are less expensive than satellites for links less than 2000 or 3000 overland miles. For coast-to-coast communications, however, the 24-hr satellite may be competitive with overland relays (Ref. 4).

XIII. Conclusions

In this paper we have arrived at the following major conclusions concerning the characteristics of a communications system using satellites.

* Cost per channel per year with the existing transatlantic cable (laid in 1956) is $240,000.

(1) A satellite system is economically and technologically feasible for point-to-point and trunk communications.

(2) The satellites should be active repeaters.

(3) The orbit should be circular with a period of either 24 or 12 hr.

(4) Operating frequencies should lie between 1 and 10 kMc/s.

(5) The power supply in the satellite should use solar cells and storage batteries.

(6) The satellites should be controlled in their orbit position.

(7) The satellites should be controlled in attitude.

(8) The satellites should have an active temperature control system.

(9) The life of the satellites will be greater than 18 months.

(10) The cumulative cost of establishing and maintaining a satellite system for 10 years will be on the order of $250 million.

REFERENCES

1. DENNEY, J. M., DOWNING, R. G., AND GRENALL, A. "High Energy Proton Damage", American Rocket Society, to be published.

2. Frequency Needs for Space Communications: Testimony and Exhibits of American Telephone and Telegraph Company before the FCC in the Matter of Allocation of Frequencies in the Bands above 890 Mc/s, 6 July 1960. Federal Communications Commission, Washington, D.C., 1960.

3. LEHR, S. N., AND TRONOLONE, V. J. "The Space Environment and Its Effects on Materials and Component Parts", presented at the Joint Reliability Seminar, Los Angeles Section, Amer. Inst. Radio Engrs., 5 December 1960.

4. MECKLING, W. "The Economic Potential of Communication Satellites", The Rand Corp., Report P-2216, 1 March 1961.

5. MARROW, W. E., Jr, AND MEYER. A. F. "Orbital Scatter Communication System", Transactions, Fourth Symposium on Ballistic Missile and Space Technology, 24–27 August 1959, Vol. 1, p. 321, WDAT 60 542.

6. MUELLER, G. E. "A Pragmatic Approach to Space Communications", Proc. Inst. Radio Engrs. 48, 558, 1960.

7. PIERCE, J. R., AND KOMPFNER, R. "Transoceanic Communications by Means of Satellites", Proc. Inst. Radio Engrs., 47, 372, 1959.

8. "Simulating Speech Through Space", Bell Lab. Rec. 38, 296, 1960.

9. SINDEN, F. W., AND MAMMEL, W. L. "Geometric Aspects of Satellite Communication', I.R.E. Transaction on Space Electronics and Telemetry, SET–6, No. 3–4, pp. 146–57, September–December 1960.

10. Statistical Abstracts of the United States, Washington, 1960.

11. Statistics of Communication Common Carriers, year ended 31 December 1958, Federal Communications Commission, Washington, D.C., 1960.

12. WHITFORD, R. K. "Design of Earth Satellite Attitude Control Systems", Report STL–2313–0001–RU–000, 24 May 1961.

THE COURIER SATELLITE

PIERCE W. SIGLIN

GEORGE SENN

U.S. Army Signal Research and Development Laboratory, Fort Monmouth, New Jersey, U.S.A.

I. INTRODUCTION

The operational applications of earth-orbiting satellites as delayed repeater radio stations which receive, store and retransmit messages upon command, offer attractive means to overcome present-day circuit congestion on existing world communications networks. One needs only to review present and projected traffic growth patterns of current systems to fully appreciate the impact of communications satellites on future network planning, and the promise held forth by future satellite-supplemented global communications networks.

Project COURIER has been a research and development experiment to demonstrate the technical and economic possibilities of a high capacity satellite communications system. The experiment has been conducted under the auspices of the Department of the Army by the U.S. Army Signal Corps and the U.S. Air Force. The U.S. Army Signal Research and Development Laboratory provided the planning, guidance and technical direction of this experiment in communications. Implementation of the programme was accomplished by prime contractors; the Philco Corporation for the satellite package, Radiation Inc. for the ground station antennas, and I.T.T. Laboratories for the ground station equipment.

The COURIER satellite provides the capability of receiving and storing messages from one ground station, and transmitting these messages at a later time upon receiving the appropriate command, to another ground station, hence the name "Courier". It is also capable of functioning as a direct real time repeater to relay voice, facsimile or teletype messages without delay, from one ground station to another ground

C.S.–G*

station, provided that both stations can "see" the satellite at the same time.

The interrelation of the ground station locations, communication service areas, duration and frequency of service provide the basic parameters from whence evolves the system design. The system includes ground station complexes which simultaneously perform the tracking, data exchange, and telemetry functions in conjunction with the satellite radio relay station. Tape recorders aboard the satellite store messages from the ground stations and, upon command, the satellite retransmits the messages to provide communication to another ground station, once acquisition of the satellite by the ground station has been accomplished and the communications links established.

The COURIER satellite carries a standard v.h.f. beacon to permit tracking by the existing world network of satellite tracking stations in order to establish ephemeris data for use by the COURIER ground stations. The ephemeris data furnished to each ground station provides a series of antenna azimuth and elevation pointing angles plotted over discrete time intervals for each pass of the satellite over a ground station. For ease of acquisition, the ground station transmits the initial TURN ON command via the v.h.f. link. With the satellite acquired, further exchanges of message traffic and commands are accomplished over the microwave link. Telemetry data are transmitted from the satellite to the ground stations over the v.h.f. link.

TABLE I. System Parameters—u.h.f. Circuits

	Satellite	Ground Station
Power output	4 W	1000 W
Noise figure	14 db	—
Noise temperature	—	640°K
i.f. Bandwidth	550 kc/s	100, 200, or 500 kc/s
Antenna gain	−4 db	41 db
Antenna polarization	Linear	Circular transmission Diversity reception
Carrier/Noise ratio at 3000 miles slant range		
Satellite-to-ground	22 db	
Ground-to-satellite	21 db	
Recorder Storage Capacity	13,200,000 bits in each of four digital recorders	
Transmission Rate	55,000 bits/sec	

PLATE 1

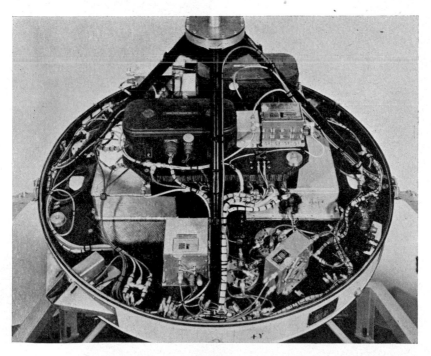

PLATE 2

PLATE 3

PLATE 4

PLATE 5

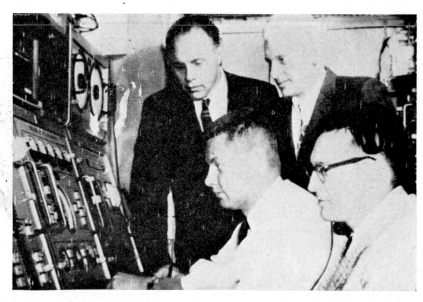

PLATE 6

The COURIER system design was based upon the satellite being injected into an approximately circular orbit at an altitude of 650 nautical miles. The significant parameters of the system are shown in Tables I and II. Carrier-to-noise ratio calculations based upon these parameters indicate reasonably strong signal paths for both v.h.f. and u.h.f. signals at slant ranges between satellite and ground station of 3000 miles.

TABLE II. System Parameters—v.h.f. Circuits

	Satellite	Ground Station
Standby power output	50 MW	—
Active power output	1·5 W	100 W
Noise figure	8 db	4 db
i.f. Bandwidth	30 kc/s	6 kc/s
Antenna gain	−4 db	19 db
Antenna polarization	Circular	Linear transmission Diversity reception
Carrier/Noise ratio at 3000 miles slant range Satellite-to-ground standby	18 db	
Satellite-to-ground active	25 db	
Ground-to-satellite	32 db	

II. Operation and Tracking

In an exchange of messages between the satellite and a ground station, the data stored on magnetic tapes is transmitted to the satellite serially at an effective 55 kilobit per second rate and stored by the satellite tape recorders assigned to other ground stations. The storage capacity of each digital recorder is 13,200,000 bits. Read-out of data from the satellite is accomplished (upon command) simultaneously with the transmission of data to the satellite. A combination of frequency diversity transmission from the satellite and frequency and polarization diversity reception by the ground station is used to minimize the transmission effects of the satellite spinning and tumbling as it speeds along in outer space. A combined output of both the frequency and polarization diversity ground station receivers is derived to feed the ground station message recovery equipment. Tracking and acquisition of the satellite are accomplished by a 28-ft dual feed, parabolic antenna.

Concentric beams at v.h.f. and u.h.f. are generated by the antenna system which comprises feed elements for both u.h.f. and v.h.f. as well as a motor-driven lens scanner to provide conical scan at u.h.f. as part of the automatic tracking antenna system. Antenna tracking accuracies

SATELLITE BLOCK DIAGRAM

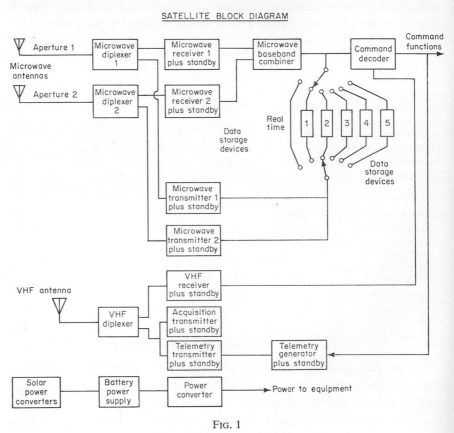

Fig. 1

of 0·5° at slew rates of 15° per second have been demonstrated. The antenna beam widths are 18° at v.h.f. and 1·35° at u.h.f.

The selection of v.h.f. operating frequencies for the acquisition, telemetry and initial command link was made to take full advantage of the wide antenna beam widths at these frequencies, which made initial acquisition of the satellite fairly simple. A microwave communications link was chosen to provide channel bandwidth compatible with the message traffic, and to take advantage of the excellent propagation and noise-free characteristics of the microwave frequencies.

Figure 1 shows a simplified block diagram of the satellite electronics. Approximately 1300 transistors and diodes plus 2 operating vacuum tubes are utilized in the relay of message traffic from one ground station to another. The v.h.f. circuitry includes dual receivers, redundant acquisition beacons, telemetry transmitters and a telemetry generator. This portion of the satellite electronics provides the tracking beacon signal, the initial TURN ON command channel, telemetry of 35 satellite parameters and acknowledgment that a command has been received. Four whip antennas, electrically fed in quadrature, generate an antenna pattern circularly polarized along an axis normal to the plane of the whips. Each v.h.f. receiver is a completely transistorized single conversion superheterodyne unit with a noise figure of 8 db. The v.h.f. beacon and telemetry transmissions are provided by separate transmitters. Redundant units are available which are switched into operation by the proper command from the ground station. The transmitters are completely transistorized. The beacon output is unmodulated c.w. and the telemetry transmitter is frequency modulated by sub-carrier oscillators to provide data on 35 satellite parameters.

The communications message traffic link between the satellite and the ground stations operates in the microwave frequency band. The 2 linearly polarized u.h.f. notch antennas provide continuous pattern coverage through a solid angle exceeding a hemisphere. Separate u.h.f. receivers with permanently connected redundant units detect the microwave signals at each antenna. Two microwave transmitters operating on frequencies spaced approximately 20 Mc/s apart provide a frequency diversity transmission to the ground station. The output of the 4 microwave receivers is combined at baseband. Each of the microwave receivers is a completely transistorized single conversion superheterodyne providing a noise figure of 14 db. The satellite carries 4 microwave transmitters, 2 primary and 2 redundant. The 2 operating transmitters are frequency modulated by the same signals. The redundant transmitter pair are switched into operation upon command from a ground station.

The recorded messages are stored on any one of the 5 magnetic tape recorder-reproducer machines. The recorders are designed to record and play back in opposite directions. For this reason traffic is read out backwards and makes necessary a compensating traffic reversal elsewhere in the system. Old traffic is automatically erased by a permanent magnet during the playback cycle. The recorders weigh 5 lb each, consume

10 W and are hermetically sealed in welded aluminium cases to insure operation of the mechanical components within a space environment.

The satellite is powered by nickel cadmium batteries which are charged by a solar cell generator. The satellite standby power between message exchange periods is 13 W; the peak power demand during full duplex message exchange is 225 W. The solar generator array made up of 19,152 1 × 2 cm solar cells supplies 60 W for charging the batteries.

A total of 21 commands may be selected by a ground station operator and transmitted to the satellite over both r.f. links. The commands which control the operation of the satellite components such as tape recorders and microwave transmitters are coded to insure privacy of operation.

Considerable emphasis has been placed on the use of redundant circuits and components in an effort to improve the overall reliability of operation. Items such as acquisition beacons, telemetry transmitters, tape recorders, and microwave transmitters are switched into the circuit by command. Other elements such as the microwave receivers, batteries and solar cell modules are permanently connected in a manner which permits continued operation of the remaining circuitry should a unit fail.

The satellite parameters which are telemetered during an operating period include temperatures, battery voltages and charging currents, power output of transmitters, a.g.c. voltages of receivers, and tape position on all tape recorders.

During a typical satellite "pass" over a ground station, the ground station antenna is directed to the point in space where the satellite should be first acquired. The sequence of operations that occur in acquiring the satellite is shown in Fig. 2. When the signal from the satellite v.h.f. beacon is received by the COURIER ground station, the station console operator issues the TURN ON command over the v.h.f. ground to satellite link. This command turns on the microwave communications transmitters and receivers in the satellite, turns off the v.h.f. beacon, activates a higher power telemetry transmitter which operates on the same v.h.f. frequency as the beacon, and switches the satellite v.h.f. receivers from a time sharing mode to one receiver operating continuously. Following the 40 sec warm-up of the microwave transmitters, the ground station antenna searches, locks on, and automatically tracks the satellite for the remainder of the pass on the microwave communications link frequency. With automatic tracking assured, an exchange of stored message information can then be simultaneously carried on between the ground station and the satellite. In a typical message exchange the ground station console operator will command a tape recorder in the satellite to "playback" recorded messages. The "playback" of a message automatically erases that message, leaving the tape recorder clear for a new message. These messages will be recorded

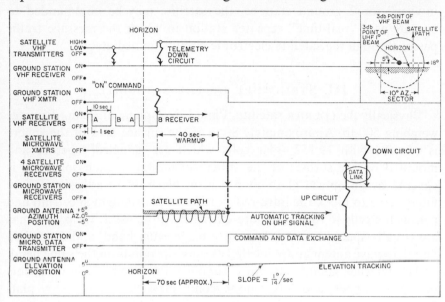

FIG. 2

on magnetic tape at the ground station. Simultaneously the ground station console operator will command another tape recorder in the satellite to "record", and messages will be transmitted to the satellite from previously recorded magnetic tapes at the ground station. This operation can be repeated for each of the 5 tape recorders carried by the satellite. Four of the recorders are used to record teletype messages in digital form, the fifth recorder is an analogue type used for recording voice or fascimile messages. The record or playback time of each recorder is 4 min. Teletype data is transmitted to or from the digital recorders at a rate of 55,000 bits per sec. In the 4 min of transmission time 13,200,000 bits, equivalent to 300,000 words can be recorded on, or played back from a digital recorder. The exchange of messages is limited by the "time-in-view" of the satellite over the ground station. Upon completion of the traffic exchange, the ground station will command the satellite to assume the standby mode of operation with only the acquisition beacon transmitting on v.h.f. Should the ground station fail to issue this standby command, the satellite will, after a 20-sec delay, automatically revert to the standby condition. This description of a typical operating sequence indicates the delayed repeater type of operation. If there is an overlap of the "time-in-view" for 2 ground stations, the satellite may be commanded to operate as a "real-time" repeater station in which the message traffic passes through the satellite electronic circuits immediately.

III. STRUCTURE OF THE SATELLITE

Physically the COURIER satellite, Plate 1, is a sphere of 52 in. diameter weighing 500 lb. Approximately 80% of the surface area of the sphere is covered with 19,152 solar cells which are shingled together and mounted on the surface of the sphere in modules. Each solar cell is covered by an optical glass filter held in place by an adhesive. These filters serve to block the infra-red portion of the sunlight spectrum and permit the cells to function more efficiently. Four v.h.f. whip antennas are equally spaced around the centre magnesium band, and 2 notched fin microwave antennas are located at opposite points on the same band.

The internal satellite structure supports the electronic components on centre, upper, and lower honeycomb glass fibre and epoxy platforms supported on aluminium tubing trusses. Plate 2 provides a top view of the satellite interior showing code tape readers, and v.h.f. receivers on the upper platform, microwave transmitters, diplexers and receivers on the centre platform. Mounted under the top platform are the acquisition and telemetry transmitters, the telemetry generator, and the message detector unit of the command decoder.

Plate 3 shows a bottom view of the satellite component arrangement with tape recorders mounted on the bottom platform, microwave transmitters and receivers plus command decoder logic boxes mounted on the under side of the centre platform. The nickel cadmium battery containers are also mounted in the middle of this central platform. Each component is thermally isolated from the others. Temperature control is achieved by the surface coatings employed on the components and the radiation geometry between the various components. The ring seen at the top of Plate 3 is the mating surface to the spin table of the missile second stage.

Plate 4 shows the microwave transmitter with associated automatic frequency control circuitry prior to hermetic sealing of the unit. Each transmitter is hermetically sealed in a welded magnesium case to insure reliable operation of the mechanical a.f.c. circuit in the environment of outer space. The rectangular box at the lower right hand side is a wax-filled heat sink which depends upon heat absorbed by the change of wax from a solid to a liquid state when heated to cool the anode of the trans-mitter tube during operation.

Plate 5 shows the completely transistorized v.h.f. telemetry trans-mitter. The frequency of this transmitter is controlled to within $\pm$ 0·002% by crystal controlled oscillators. The power output is 1 W.

All satellite components were tested on a vibration machine at a vibration level of 11·5 g over a white noise random spectrum of 20–2000 c/s for a 3 min period. The complete satellite was tested at a level of 7·5 g over the same white noise bandwidth. COURIER satellites were also tested in vacuum chambers under simulated orbital environmental con-ditions. In the installation used for the COURIER, liquid nitrogen was circulated through the chamber walls to simulate the cold of outer space. The pressure within the chamber was reduced to 10^{-5} mm Hg. Simulation of heat from the sun was provided by an electrical heating element which would be moved around the satellite at a speed corres-ponding to the satellite spin rate.

IV. GROUND STATIONS

The ground stations were located at Fort Monmouth, New Jersey, and at Camp Salinas, Puerto Rico. These locations permitted experi-ments to be conducted using the satellite in both delayed repeater as well as the real time mode of operation. The altitude and inclination of the satellite orbit made possible an average of 5 workable orbits at the Fort Monmouth station and an average of 7 at the Puerto Rico station out of approximately 14 orbits per day. From 8 to 19 min of operating time was available to the ground stations for these usable orbits with considerable overlap in the time in view of the 2 stations. This over-lap permitted experiments to be conducted in the real time relay of message traffic.

The complete station at Fort Monmouth, New Jersey, includes a 28-ft parabolic antenna, 3 semi-trailers and a maintenance van. The semi-trailers provide for communications co-ordination control, operating console and message processing centre, and a radio van containing all the u.h.f. and v.h.f. receiving and transmitting equipment. A servicing tower for the antenna feed is situated behind the 40-ft antenna tower.

A simplified block diagram for the entire ground station is shown in Fig. 3. The antenna feed comprises feed elements for both u.h.f. and

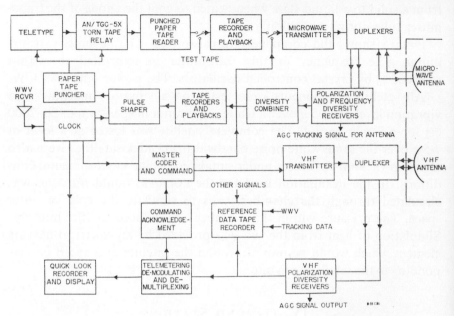

Fig. 3

v.h.f. as well as a motor-driven lens scanner to provide conical scan at u.h.f. for the automatic tracking antenna system. In the v.h.f. portion of the ground station, polarization diversity reception is utilized with switching provided to permit either horizontal or vertical linear transmission. The satellite acquisition beacon and telemetry signals are received over this v.h.f. equipment. The telemetry information is recorded on magnetic tape and on a strip-chart recorder. The strip-chart record is used to show the tape recorder loading situation and the charge condition of the satellite nickel–cadmium batteries at the beginning of an operating period.

The u.h.f. system uses circular polarization on transmission and both frequency and polarization diversity in reception. The ground station receiving system utilizes 4 parametric converters in a double conversion superheterodyne system to provide both frequency and polarization diversity reception. The non-coherent transmission of the 2 satellite transmitters requires that both pre-detection and post-detection signal combining be used in the receivers. The receiving system has a dynamic range of -140 to -100 dbW.

A speed buffering technique is used to attain the high teletype message traffic capacity of the COURIER system. Teletype paper tape messages are converted into electrical signals at a speed of 200 characters per sec and recorded on magnetic tape moving at a speed of 1·68 in per sec. Each character of the message is recorded in parallel on 6 tracks of the magnetic tape. The magnetic tape is then played back at a tape speed of 60 in per sec. The signals from the magnetic tape are converted from parallel to serial form and used to frequency modulate the microwave signal to the satellite at a rate of 55,000 bits per sec. The seventh or stop bit for each character is inserted in the parallel to serial converter. Incoming traffic from the satellite to the ground station microwave receivers is converted from serial to parallel form by a converter and recorded on a magnetic tape at a tape speed of 60 in per sec. The tape is then played back at a 1·68 in per sec rate to a paper tape punch operating at a speed of 200 characters per sec. The resulting paper tape can be used by standard teletype machines.

Recorded voice traffic is handled in a different manner. The analogue message is initially recorded on magnetic tape at unity speed. The tape is then reversed and played back to the satellite at twice unity speed. Analogue traffic received from the satellite is recorded on magnetic tape at twice unity speed. When played back at unity speed the original message will be recovered.

V. THE COURIER 1B LAUNCHING

On 4 October 1960 at 1250 E.S.T. the COURIER 1B satellite was launched from Cape Canaveral, Florida. The launch was due east from pad 17B at Cape Canaveral. The launch sequence placed the COURIER satellite into an orbit inclined 28·3° to the earth's equator. The actual injection of the satellite into orbit occurred over South Africa, as shown in Fig. 4.

The complete THOR-ABLE-STAR missile with satellite payload stands over 79 ft tall and weighs more than 105,000 lb at lift-off.

Figure 5 shows the predicted path of the first COURIER 1B orbit. The figures denote the number of minutes after the injection into orbit.

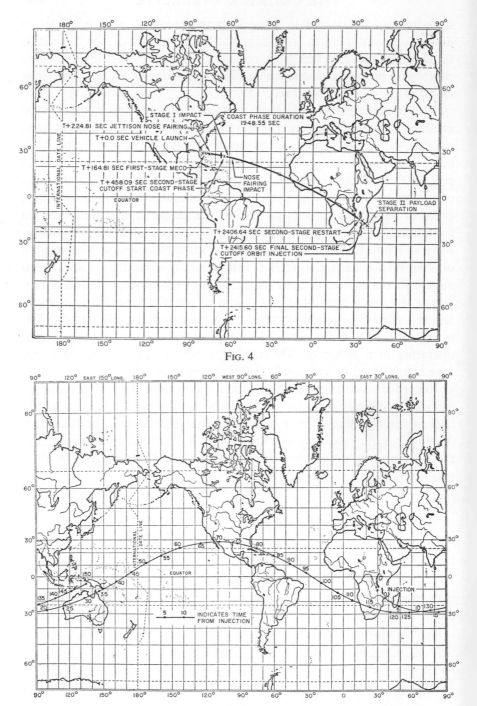

Fig. 4

Fig. 5

The launch was completely successful and attained a satellite orbit with an apogee of 755 miles, a perigee of 598 miles and an orbit period of 107 min.

The communications system far exceeded the expectations of the designers. The received power at the ground stations has averaged above —95 dbm. Experiments in handling both delayed repeater and real time teletype and voice traffic have been successful. Plate 6 is a copy of a photograph received over the COURIER satellite on 12 October 1960 via real time facsimile transmission. The total path length in this facsimile transmission was over 4000 miles. The few white specks in the photograph are noise. The photograph shows the operating console of the Fort Monmouth, New Jersey, Courier Station. From left to right standing in the background are the Programme Director, Mr George Senn, and the Programme Manager, Mr Pierce W. Siglin. Seated in the foreground are the Station Manager, Mr Samuel Findler, and the Station Engineer, Mr Walter Teetsel.

Unfortunately, since 22 October 1960 the satellite has not responded to any commands transmitted from the ground stations. The acquisition beacon continues to transmit indicating that the satellite is in the standby mode of operation.

Although the operating period has been short, the Courier satellite communications system has demonstrated a capacity competitive with transatlantic h.f.r. and cable links. With the refinement of techniques it may be assumed that the system can be made economically competitive.

SUBJECT INDEX

A

Acceleration, 89
Altitude, *see* Orbits, height
Amplifiers, *see also* Receivers, 82–83, 167, 168, 172
   maser, 25, 37, 38, 49, 91, 92, 93, 172
   travelling wave tubes, 167, 168, 169
Antennae
   ground, 10, 39–40, 49, 87, 91–93, 109, 137, 171, 172, 186, 187, 188, 194
   satellite, 8, 41, 48, 49, 55, 64, 92, 123, 126, 128, 137, 150, 156, 157, 160, 167, 169, 171, 175, 186, 187, 189, 192
Atmosphere, density, 151
Attenuation
   due to rain, 39, 45–46
   ionospheric, 40–41
   tropospheric, 39, 42
Attitude stabilization, 8, 90, 91, 109, 160–164
Audio frequency, 31, 34, 48, 49

B

Bands, *see also* Radio frequency, 10, 18, 19, 21, 24–29, 31, 34–38, 50, 53, 54, 56–60, 62, 64–65, 66, 92, 93, 122, 126, 129, 135, 137, 170, 189
   allocation, 41–45
   assembly, definition, 69
   availability, 109–110
   band compression, 60
   information band, 34–35, 36, 38, 50, 51, 52, 57, 110, 134, 136, 138, 174, 186, 187
   Nyquist band, 34, 35, 53, 58
   sidebands, 24, 28, 36, 37
   single, 48, 49, 50, 53, 56, 63, 65, 68
   vestigial, 53, 57, 60, 69
   video, 60
Batteries, 159
Beacons, 186, 192

C

Cables, 31, 86, 95, 96, 111, 133, 134, 182
Cape Canaveral, 112, 195
Christmas Island, 5
Circuits, 2-wire, 4-wire, 17, 20, 21, 22, 23, 67, 87, 120
Coding and decoding, *see also* Modulation
   beam coding tubes, 75
   binary, 24, 32–35, 48, 52, 53, 55, 59, 60, 70–73, 75, 76, 77, 80
   decoding, 80–82
   equal probability code, 70–73
   power economy code, 33–34, 54–56, 58, 60, 63, 64, 73
Colomb Bechar, 4
Command, ground, 89, 90, 173–176, 186, 189, 190, 192
Commonwealth service, 95–112
Communications satellite projects
   Courier, 10, 185–197
   Echo, 85, 87, 91, 92, 94, 139
   "Needles", 138, 139
Communications satellites, *see also* Costs; Telecommunications,
   active, 7, 8, 12, 13, 17, 39–40, 85–89, 94, 109, 113, 115, 117–119, 122–123, 138–141, 150, 155, 167–183, 185–197
   active broadcast, 113, 123–129
   angle of elevation, 46–48, 49, 143
   Courier, 185–197
   ground track, 104, 105, 106, 107, 144
   life, *see* Reliability
   manned, 126, 128
   military, 8, 132, 134, 137, 185–197
   number, *see* Coverage
   passive, 7, 12, 17, 25, 38–39, 94, 138–141
   range, 46–48, 49, 50, 51, 54, 56, 64
   replacement cost, 178
   replacement time, 118, 119
   station keeping, *see also* Attitude stabilization, 90, 109